Lecture Notes in Mathematics

Edited by A. Dold and B. Eckmann

757

Smoothing Techniques for Curve Estimation

Proceedings of a Workshop held in Heidelberg, April 2 – 4, 1979

Edited by
Th. Gasser and M. Rosenblatt

Springer-Verlag
Berlin Heidelberg New York 1979

Editors

Th. Gasser
Dept. of Biostatistics
Central Institute of Mental Health
J 5, P.O. Box 5970
D-6800 Mannheim

M. Rosenblatt
Dept. of Mathematics
University of California
San Diego, La Jolla
California 92032
USA

AMS Subject Classifications (1970): 62 G05, 62 G20, 65 D07, 65 D10

ISBN 3-540-09706-6 Springer-Verlag Berlin Heidelberg New York
ISBN 0-387-09706-6 Springer-Verlag New York Heidelberg Berlin

Library of Congress Cataloging in Publication Data
Main entry under title:
Smoothing techniques for curve estimation.
(Lecture notes in mathematics; 757)
"The workshop . . . has taken place as part of the activities of the Sonderforschungsbereich 123, "Stochastic Mathematical Models."
Bibliography: p.
Includes index.
1. Estimation theory--Congresses. 2. Curve fitting--Congresses. I. Gasser, Theodor A., 1941- II. Rosenblatt, Murray. III. Series: Lecture notes in mathematics (Berlin); 757.
QA3.L28 no. 757 [QA276.8] 510'.8s [519.5'4] 79-22814
ISBN 0-387-09706-6

Printing and binding: Beltz Offsetdruck, Hemsbach/Bergstr.
2141/3140-543210

PREFACE

The workshop 'Smoothing Techniques for Curve Estimation' has taken place as part of the activities of the Sonderforschungsbereich 123, "Stochastic Mathematical Models". The participants and the organizer agree that it was a lively and successful meeting. Our thanks go to the Deutsche Forschungsgemeinschaft for enabling this meeting as part of the visiting program of the Sonderforschungsbereich 123. Our hope is that the ties founded or strengthened during the meeting will continue to be fruitful.

Heidelberg, July 1979

TABLE OF CONTENTS

NONPARAMETRIC CURVE ESTIMATION:

Some Introductory Remarks

Th. Gasser
Zentralinstitut für Seelische Gesundheit
Abteilung Biostatistik
Postfach 5970
D-6800 Mannheim 1

M. Rosenblatt
University of California, San Diego
La Jolla, California 92032/USA

The workshop on smoothing techniques for curve estimation was organized because of the increasing theoretical and applied interest in such questions. Making a histogram of a data set is a time-honored tool in statistics as well as in other areas. The notion of representing and smoothing data in more flexible ways arises naturally. Given any way of representing a function, one can adapt such a representation to give a method of smoothing data. In this way one obtains attractive alternatives to a parametric analysis. There is interest in nonparametric curve estimation because parametric models require assumptions which are often unwarranted and not checked when entering a new field in the empirical sciences. The availability of computer equipment, especially with graphics terminals, allow one to avoid undue assumptions and "let the data speak for themselves".

We first make some historical remarks. One of the earliest papers to suggest using kernel estimates of density functions was that of Rosenblatt (1956). A few years later on further results on the large sample behavior of such estimates were obtained in Bartlett (1963) and Parzen (1962). Estimates making use of a Fourier representation were suggested in a paper of Censov (1962). These techniques have been used by Tarter and Raman (1972) in a biomedical context. Nearest neighbor estimates have been proposed for density estimation in Loftsgaarden and Quesenberry (1965), and for regression analysis in Stone (1977). In a recent paper of Mack and Rosenblatt (1979), the large sample behavior of nearest neighbor density estimates is determined. Spline methods have also been of considerable interest and were proposed as a basis for density estimation in Boneva, Kendall and Stefanov (1971). The large sample properties of cubic spline density estimates are discussed in the paper of Lii and Rosenblatt (1975). Both spline and kernel estimates are used to estimate the probability density of the derivative of turbulent velocity readings. This is of interest in a modified model of Kolmogorov where it is suggested that this probability density should be approximately log normal. This is not consistent with the analysis out in the tails (say beyond three sigma) of the distribution. Spline estimates have been used in an analysis arising in an archaeological context in Kendall (1974). The use of nonparametric (say kernel) regression estimates to deal with the important problem of calibrating radiocarbon and bristlecone pine dates has been proposed in Clark (1974). The use of kernel estimates for discriminant analysis in a multidimensional context has been examined by Van Ness and Simpson (1976) and evaluated favorably.

One of the important questions when dealing with such nonparametric estimates is that relating to the choice of bandwidth or the degree of smoothing. Papers of Rosenblatt (1971) and Bickel and Rosenblatt (1973) have considered global behavior and global measures of deviation of kernel density estimates. An elegant and powerful result of Komlos et al (1975) is very useful in obtaining such results. Silverman (1978, 1979) has used related ideas and suggested an interesting way of choosing a bandwidth based on them. A number of applications are discussed in Silverman (1979).

We should now like to mention another area in which nonparametric curve estimates have been useful, that of growth and development. The analysis of the Zürich longitudinal growth study (Stützle, 1977; Largo et al 1978, Stützle et al 1979) is a case in point. The analysis started with the classical problem of height growth, and in particular of the pubertal growth spurt. Polynomial models were recognized as unsuitable. The logistic and the Gompertz functions have been used and compared (Marubini et al, 1971), but do not fit well over the whole period of growth. This limits the span of interpretation seriously. Bock et al (1973) have proposed a double logistic model between 1 and 18 years of age (addition of two logistic functions, associated with pubertal and prepubertal growth respectively). The fit is not good, and there is an age-dependent bias; the model has also led to qualitatively dissatisfying 'facts': It suggests that the difference between boys and girls resides primarily in the prepubertal parameters, contrary to everyday experience, and that the pubertal component starts in early childhood. Preece and Baines

(1978) recently introduced a parametric family which gives a better fit, as measured by the residual sum of squares.

Smoothing procedures offer an alternative for obtaining a set of interpretative parameters. Tanner et al (1966) carried out smoothing by eye, a procedure which is time-consuming, not reproducible and biased. (Tanner et al, 1966) had a too accentuated pubertal spurt). The bias of spline or kernel smoothing depends in a simple way on the function to be estimated. In the Zürich growth study, cubic smoothing splines (following Reinsch, 1967) have been used (Largo et al, 1978). The choice of the smoothing parameter is critical and should ideally be determined from the data. A cross-validation procedure suggested by Wahba and Wold (1975) gave in general good results. A general feature of splines smoothing is the relatively high cost in computer time and/or core (particularly annoying with large data sets encountered in neurophysiology). This draws our attention to alternatives, as e.g. kernel estimates.

<u>REFERENCES:</u>

Bartlett, M.S. (1963): Statistical estimation of density functions. Sankhya Sec. A 25 245-254

Boneva, L.I., Kendall, D.G., and Stefanov, I. (1971): Spline transformations, J. Roy. Statist. Soc. B. 33, 1-70

Bickel, P.J. and Rosenblatt (1973): On some global measures of the deviations of density function estimates. Ann. Statist. 1, 1071-95

Bock, R.D., Wainer, H. Petersen, A., Thissen, D., Murray, J., Roche, A. (1973): A parametrization for individual human growth curves, Human Biology 45, 63-80

Censov, N.N. (1962): Evaluation of an unknown distribution density from observations. Soviet Math. 3, 1559-1562

Clark, R.M. (1974): A survey of statistical problems in archaeological dating. J. Multiv. Anal. 4, 308-326

Kendall, D.G. (1974): Hunting quanta. Phil. Trans. Roy. Soc. London, A 276, 231-266

Komlos, J., Major, P. and Tusnady, G. (1975): An approximation of partial sums of independent random variables. Zeit. für Wahr. 32, 111-131

Largo, R.H., Stützle, W., Gasser, T., Huber, P.J., Prader, A. (1978): A description of the adolescent growth spurt using smoothing spline functions. Annals of Human Biology, in print.

Lawton, W.H., Sylvestre, E.A., Maggio, M.S. (1972): Self modeling nonlinear regression. Technometrics 14, 513-532

Lii, K.S., Rosenblatt, M. (1975): Asymptotic behavior of a spline estimate of a density function. Comput. Math. Appl. 1, 223-235

Mack, Y.P. and Rosenblatt, M. (1979): Multivariate k-nearest neighbor density estimates. J. Multiv. Anal. 9, 1-15

Marubini, E., Resele, L.F., Barghini, G. (1971): A comparative fitting of gompertz and logistic functions to longitudinal height data during adolescense in girls. Human Biology 43, 237-252

Parzen, E. (1962): On the estimation of a probability density and mode. Ann. Math. Statist. 33, 1065-1076

Preece, M.A., Baines, M.J. (1978): A new family of mathematical models describing the human growth curve. Annals of Human Biology, 5, 1-24

Reinsch, Ch. (1967): Smoothing by spline functions. Num. Math. 10, 177-183

Rosenblatt, M. (1956): Remarks on some nonparametric estimates of a density function. Ann. Math. Statist. 27, 832-837

Rosenblatt, M. (1970): Density estimates and Markov sequences. In Nonparametric Techniques in Statistical Inference, M. Puri ed. 199-210

Rosenblatt, M. (1971): Curve estimates. Ann. Math. Statist. 42, 1815-1842

Silverman, B.W. (1978): Choosing a window width when estimating a density. Biometrika 65, 1-11

Silverman, B.W. (1979): Density estimation: are theoretical results useful in practice? - presented at a meeting in honor of W. Hoeffding

Stützle, W. (1977): Estimation and parametrization of growth curves. Thesis 6039 ETH Zürich

Stützle, W., Gasser, Th., Largo, R., Huber, P.J., Prader, A., Molinari, L. (1979): Shape-invariant modeling of human growth. Mimeographed manuscript, 1979

Tanner, J.M., Whitehouse, R.H., Takaishi, M. (1966): Standards from birth to maturity for height, weight, height velocity and weight velocity: British Children. Archives of Disease in Childhood 41, 451-471, 613-635

Tarter, M. and Raman, S. (1972): A systematic approach to graphical methods in biometry. Proceedings of 6th Berkeley Symposium vol. IV, 199-222

Van Ness, J.W. and Simpson, C. (1976): On the effects of dimension in discriminant analysis. Technometrics 18, 175-187

Wahba, G., Wold, S. (1975): A completely automatic French curve: Fitting spline functions by cross-validation. Communications in statistics 4, 1-17

A TREE-STRUCTURED APPROACH TO NONPARAMETRIC MULTIPLE REGRESSION

Jerome H. Friedman*
Stanford Linear Accelerator Center
Stanford, California 94305/USA

Introduction

In the nonparametric regression problem, one is given a set of vector valued variables $\underline{X}$ (termed carriers) and with each an associated scalar quantity Y (termed the response). This set of carriers and associated responses $\{Y_i,\underline{X}_i\}$ $(1\le i\le N)$ is termed the training sample. In addition (usually at some later time), one is given another set of vector valued variables $\{\underline{Z}_j\}$ $(1\le j\le M)$ without corresponding responses and the problem is to estimate each corresponding response using the values of its carriers and the training sample. That is:

$$\hat{Y}(\underline{Z}_j) = \text{Rule}\ [\underline{Z}_j,\ \{Y_i,\underline{X}_i\}\ (1 \le i \le N)]\ (1 \le j \le M).$$

The rule for performing the estimation is usually referred to as the model or regression function.

In addition to this basic predictive role, there are usually other data analytic goals. One would like the model to reveal the nature of the dependence of the response on the respective carriers and lend itself to easy interpretation in a similar manner to the way parametric models often do via the fitted values of their parameters.

Binary Regression Tree

The nonparametric regression models discussed herein are based on binary trees. A binary tree is a rooted tree in which every node has either two sons (nonterminal nodes) or zero sons (terminal nodes). Figure 1 illustrates a simple binary tree.

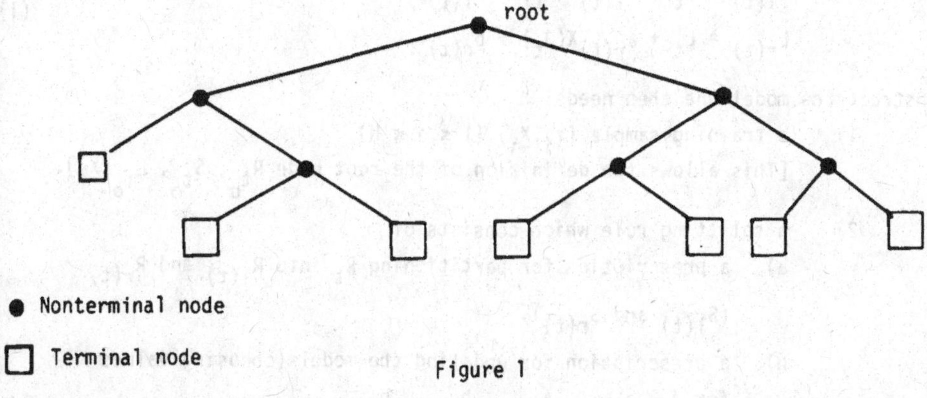

Figure 1

*This work is part of a joint research effort by Leo Breiman, Jerome Friedman, Lawrence Rafksy and Charles Stone. Work partially supported by the Department of Energy under contract number EY-76-C-03-0515.

For these models, each node t represents:

1) a subsample S_t of the training sample,
2) a subregion R_t of the carrier data space,
3) a linear model $L_t(\underline{X}) = \underline{A}_t \cdot \underline{X} + B_t$ to be applied to $\underline{X} \in R_t$.

(For the models discussed in this report, the subsample S_t, represented by node t, is just the set of training vectors that lie in its corresponding subregion R_t.)

In addition, each <u>nonterminal</u> node represents:

4) a partitioning or splitting of R_t into two disjoint subregions $R_{l(t)}$ and $R_{r(t)}$

$(R_{l}(t) \cup R_{r(t)} = R_t$ and $R_{l(t)} \cap R_{r(t)} = 0)$

and a corresponding partitioning of S_t into two disjoint subsets $S_{l(t)}$ and $S_{r(t)}$.

The binary regression tree is defined recursively: let t_o be the root node and

S_{t_o} = entire training sample

R_{t_o} = entire carrier data space

$L_{t_o}(\underline{X})$ = linear (least squares fit) of Y on $\underline{X}$ using S_{t_o}.

Let t be a nonterminal node with left and right sons l(t) and r(t) respectively. Then

$R_{l(t)}$ and $R_{r(t)}$ are the subregions defined by the partitioning of t,

$S_{l(t)}$ and $S_{r(t)}$ are the subsamples defined by the partitioning of t.

The linear models associated with the left and right sons are derived from the parent model by modifying the dependence on one of the carriers J_t:

$$
\begin{aligned}
L_{l(t)} &= L_t + a_{l(t)} X(J_t) + b_{l(t)} \\
L_{r(t)} &= L_t + a_{r(t)} X(J_t) + b_{r(t)}.
\end{aligned}
\tag{1}
$$

To construct the model one then needs:

1) a training sample $\{Y_i, \underline{X}_i\}$ $(1 \le i \le N)$
[This allows the definition of the root node R_{t_o}, S_{t_o}, $L_{t_o}(\underline{X})$],

2) a splitting rule which consists of
   a) a prescription for partitioning R_t into $R_{l(t)}$ and $R_{r(t)}$ ($S_{l(t)}$ and $S_{r(t)}$),
   b) a prescription for updating the model (choosing values for J_t, $a_{l(t)}$, $a_{r(t)}$, $b_{l(t)}$, $b_{r(t)}$, to get $L_{l(t)}$ and $L_{r(t)}$
   (thereby defining the two son nodes of t),

3) stopping (termination) rule for deciding when not to split a node, thereby making it a terminal node.

Splitting Rule

The situation at a node that is to be split is depicted in Figure 2.

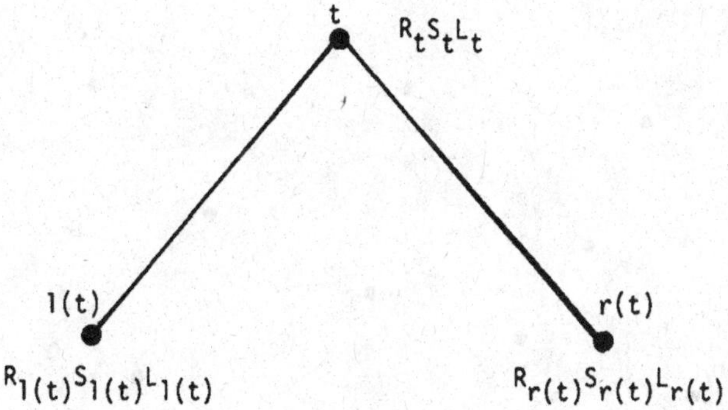

Figure 2

One has the subregion (subsample) and model associated with the parent [R_t (S_t) and L_t] and one would like to define the corresponding quantities for the two sons so as to best improve the fit of the model to the training sample. Let

$$\hat{Q}_t = \sum_{i \in S_t} [Y_i - L_t(\underline{X}_i)]^2 \tag{2}$$

be the empirical residual sum of squares associated with the parent and $\hat{Q}_{l(t)}$ and $\hat{Q}_{r(t)}$ be the corresponding quantities for the two sons. Then

$$\hat{I}_t = \hat{Q}_t - \hat{Q}_{l(t)} - \hat{Q}_{r(t)} \tag{3}$$

is an estimate of the improvement as a result of splitting node t. A reasonable goal is then to choose the partitioning so as to maximize $\hat{I}_t$ subject to possible limitations such as continuity and computability.

Since $L_{l(t)}$ and $L_{r(t)}$ are linear models (on $R_{l(t)}$ and $R_{r(t)}$) and $R_{l(t)} \cup R_{r(t)} = R_t$, one can think of [$L_{l(t)}$, $L_{r(t)}$] as a piecewise-linear model on R_t. From (1)

$$\begin{aligned} L_{l(t)} - L_t &= a_{l(t)} X(J_t) + b_{l(t)} \\ L_{r(t)} - L_t &= a_{r(t)} X(J_t) + b_{r(t)} \end{aligned} \tag{4}$$

so that we want to choose the parameters on the RHS of (4) to best fit the residuals

$$r_i = Y_i - L_t(\underline{X}_i) \qquad (i \in S_i) \qquad (5)$$

to the model associated with the parent node.

Consider the residuals (5) as a function of each of the carriers X(j) in turn. If $L_t(\underline{X})$ provides an adequate description of the dependence of the response on X(j), then there should be little structure in the values of the residuals when ordered on X(j). That is, a plot of r versus X(j) would resemble that of Figure 3a.

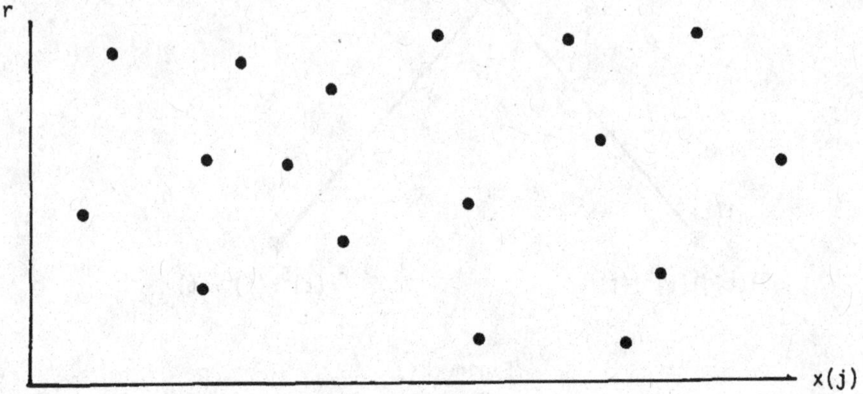

Figure 3a

On the other hand, considerable structure in the residuals (e.g., Figure 3b) would indicate that $L_t(\underline{X})$ does not provide an adequate description of the dependence of the response on X(j).

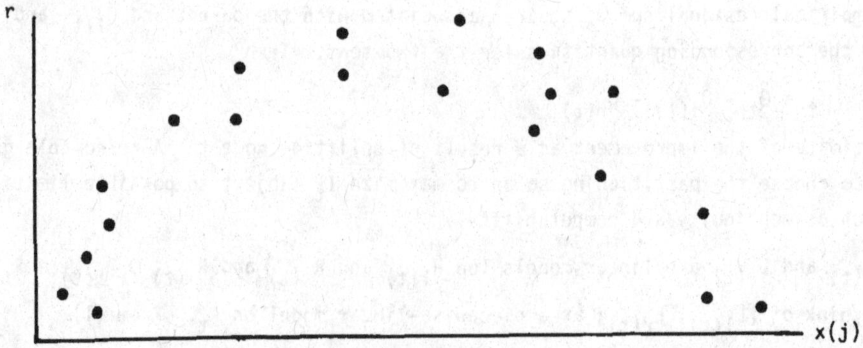

Figure 3b

The example of Figure 3b indicates a possible quadratic dependence of the residuals (and hence the response) on carrier X(j).

These observations motivate our splitting procedure. Each carrier X(j) is considered in turn. For each, a (univariate) continuous piecewise linear model is fit to the residuals from $L_t(\underline{X})$. That is, the model

$$\begin{aligned} r &= a_{lj}\,[X(j) - s_j] + b_j \qquad X(j) \le s_j \\ r &= a_{rj}\,[X(j) - s_j] + b_j \qquad X(j) > s_j \end{aligned} \tag{6}$$

is fit to $\{r_i, X_i(j)\}$ $(i \in S_t)$ by minimizing

$$Q_j = \sum_{i=1}^{k} [r_i - a_{lj}(X_i(j) - s_j) - b_j]^2 + \sum_{i=k+1}^{\#S_t} [r_i - a_{rj}(X_i(j) - s_j) - b_j]^2 \tag{7}$$

with respect to j, a_{lj}, a_{rj}, b_j, and s_j. Here the $X_i(j)$ are ordered in ascending value and $X_k(j) \le s_j$ and $X_{k+1}(j) > s_j$. That is, the best (in the least squares sense) continuous piecewise linear fit (with s_j as the knot) is made to the residuals versus each carrier X(j) and the best fit (over the carriers) is chosen.

Let the optimum values found for j, a_{lj}, a_{rj}, b_j, s_j be represented by J, a_l, a_r, b and s respectively. These solution values are used to both define the partitioning and update the model:

$$\begin{aligned} &\text{For } \underline{X} \in R_t: \\ &\qquad \text{If } X(J) \le s, \text{ then } \underline{X} \in R_{l(t)} \\ &\qquad \text{If } X(j) > s, \text{ then } \underline{X} \in R_{r(t)} \\ &\qquad L_{l(t)}(\underline{X}) = L_t(X) + a_l\,[X(J) - s] + b \\ &\qquad L_{r(t)}(\underline{X}) = L_t(\underline{X}) + a_r\,[X(J) - s] + b\,. \end{aligned} \tag{8}$$

If the model associated with the parent node is

$$L_t(\underline{X}) = \sum_{j=1}^{p} A_t(j)\,X(j) + B_t$$

then from (8), the corresponding quantities for the son nodes are:

$$\begin{aligned} A_{l(t)}(j) &= A_{r(t)}(j) = A_t(j) \qquad j \neq J \\ A_{l(t)}(J) &= A_t(J) + a_l \\ A_{r(t)}(J) &= A_t(J) + a_r \\ B_{l(t)} &= B_t - a_l\, s + b \\ B_{r(t)} &= B_t - a_r\, s + b. \end{aligned} \tag{9}$$

Thus, the models associated with the left and right sons differ from the parent and each other only in their dependence on carrier J, and the constant terms are adjusted for continuity at the split point s.

After the split is made and the model updated for the two son nodes, the above procedure is applied recursively to l(t) and r(t) and their sons and so on until the nodes meet a terminal condition. This stops the splitting making terminal nodes. Starting with the root, this recursive procedure then defines the entire regression tree.

Stopping (Termination) Rule

The recursive splitting described above cannot continue indefinitely. At some point, the cardinality of the subsample $\#(S_t)$ will be too small to reliably estimate the parameters for defining the splitting and updating the model. Thus, a sufficient condition for making a node terminal is that the size of its subsample is too small to continue splitting.

Using this condition as the sole one for termination, however, can cause serious overfitting. Basically, a split should not be made if it is not worthwhile. That is, it does not improve the model fit. The quantity $\hat{I}_t$ (3) is an estimate of the improvement in the fit as a result of splitting node t. This quantity is always positive, indicating that the empirical residual sum of squares will always improve as a result of choosing the optimum splitting. However, since the empirical residual sum of squares is an optimistically biased estimate of the true residual sum of squares from the model, a positive value for $\hat{I}_t$ does not guarantee a positive value for the true improvement I_t. A more reasonable criterion would be:

If $\hat{I}_t > k$ accept split at t and continue, otherwise make t a terminal node.

The quantity k is a parameter of the procedure, the interpretation of which is discussed below. Although lack of sufficient fit improvement (as estimated by $\hat{I}_t$) is a necessary condition for making t a terminal node, it is not sufficient. It is possible that a particular split, although not yielding much improvement itself, can make it possible for further splitting to make dramatic improvements. This would be the case, for example, if there were substantial interaction effects between pairs or sets of carriers. A sufficient condition for making a node terminal would be if its split and all further splits of its descendants yield insufficient empirical improvement. This is illustrated in Figure 4.

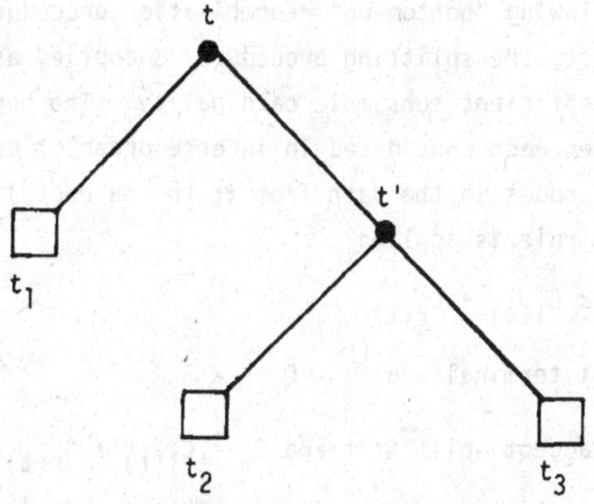

Figure 4

Here node t is split forming son nodes t_1 (which subsequently becomes terminal) and t'. Right son t' is further split forming nodes t_2 and t_3 which become terminal. The improvement associated with node t (and all further splits) is then defined to be

$$\hat{I}_t = \hat{Q}_t - \hat{Q}_{t_1} - \hat{Q}_{t_2} - \hat{Q}_{t_3} \tag{10}$$

That is the difference between the empirical residual sum of squares at node t and the sum of those associated with all terminal descendants of t. A reasonable condition for making t a terminal node is then

If $\hat{I}_t \leq 2\,k$ make t terminal, otherwise accept split at t. (11)

The factor of two on the RHS of the inequality comes from the fact that two splits were required to form these three terminal nodes and this introduces even more optimistic bias than just one split. The condition (11) can be rewritten

If $\hat{Q}_t + k \leq \hat{Q}_{t_1} + k + \hat{Q}_{t_2} + k + \hat{Q}_{t_3} + k$ (12)

make t terminal,

Otherwise, accept split at t.

This suggests associating a cost C_t with each node t of the tree as follows:

If t is terminal $C_t = \hat{Q}_t + k$

If t is nonterminal $C_t = \sum_{i \in t} C_{t_i}$ (13)

where the summation is over all terminal descendants of t. The decision to make a node terminal or not is then taken so as to minimize this cost. Note that if both sons of t [l(t) and r(t)] are terminated according to this prescription, then

$$\sum_{i \in t} C_{t_i} = C_{l(t)} + C_{r(t)} \,. \tag{14}$$

This suggests the following "bottom-up" recombination procedure for terminating the regression tree. First, the splitting procedure is applied as far as possible, terminating only for insufficient subsample cardinality. The nonterminal nodes of the resulting tree are then each considered in inverse order of depth. (The depth of a node is the number of nodes in the path from it to the root.) At each such node, the following termination rule is applied:

$$\underline{\text{If}}\ \hat{Q}_t + k \le C_{l(t)} + C_{r(t)}$$

$$\underline{\text{then}}\ \text{make } t \text{ terminal and } C_t = \hat{Q}_t + k \tag{15}$$

$$\text{Otherwise accept split at } t \text{ and } C_t = C_{l(t)} + C_{r(t)}.$$

This bottom-up recombination procedure insures that a node is made terminal only if its splitting and all possible further splitting yields insufficient improvement to the fit of the model, as determined by the improvement threshold parameter k.

This bottom-up recombination algorithm can be more easily understood intuitively by considering the following optimization problem. Let $\mathcal{T}$ be the set of all possible trees obtained by <u>arbitrarily</u> terminating the splitting procedure of the previous section. Let $T \in \mathcal{T}$ be one such tree and define its size $|T|$ to be the number of its terminal nodes. Let $\hat{Q}(T)$ be the empirical residual sum of squares associated with the regression model defined by T. The optimization problem is to choose that tree $T_k \in \mathcal{T}$, such that $\hat{Q}(T_k) + k|T|$ is minimum (breaking ties by minimizing $|T|$). The quantity k is a positive constant called the complexity parameter and T_k is said to be the optimally terminated tree for complexity parameter k. The complexity parameter is the analogue for this procedure to the smoothness parameter associated with smoothing splines or the bandwidth parameter associated with kernal estimates. Since $\hat{Q}(T_k)$ is monotone decreasing with increasing $|T_k|$, the value of k limits the size of the resulting optimally terminated tree T_k. Larger values of k result in smaller trees.

It can be shown (Breiman and Stone, 1977) that the bottom-up recombination procedure described above is an algorithm for solving this optimization problem where the complexity parameter k is just the improvement threshold parameter of that procedure. Thus, although motivated heuristically, that procedure is seen to have a natural interpretation in terms of generating optimally terminated trees T_k.

The complexity parameter k is the only parameter associated with this model. Ideally, its value should be chosen to minimize the true residual sum of squares $Q(T_k)$ associated with the model. This quantity is, of course, unavailable since only the training sample is provided. One could apply crossvalidation (e.g., see Breiman and Stone, 1977) or bootstrapping (Efron, 1977) techniques to obtain a less biased estimate of $Q(T_k)$ than $\hat{Q}(T_k)$. These estimates could be performed for various values of k and the best one chosen based on those estimates. However, this procedure is quite expensive computationally and not always reliable. Fortunately, a simple graphical procedure allows one to obtain a reasonable estimate for a good value of the complexity parameter.

It can be shown (see Breiman and Stone, 1977) that for $k' > k$, $T_{k'}$ is a subtree of T_k (i.e., $T_{k'} \subset T_k$). To obtain $T_{k'}$, one simply applies the bottom up recombination procedure to T_k using the value k'. For $k' >> k$, $T_{k'}$ will likely be much smaller than T_k, while for k' only slightly larger than k the two trees will probably be identical. One can determine the smallest value of k' that will cause $T_{k'}$ to be smaller than T_k. For each nonterminal node $t \in T_k$, one has from (15)

$$\hat{Q}_t + k > \sum_{i \in t} (\hat{Q}_{t_i} + k) \tag{16}$$

where the summation is over all terminal descendants of t. If this were not the case, the node t would have been terminal in T_k. One can associate with each nonterminal node the complexity parameter value $\bar{k}_t$ that would cause it to become terminal. From (16) one has

$$\bar{k}_t = \frac{\hat{Q}_t - \sum_{i \in t} \hat{Q}_{t_i}}{|t| - 1} \tag{17}$$

where $|t|$ is the number of terminal descendants of t. The minimum value of $\bar{k}_t$ over all nonterminal nodes of T_k is the smallest complexity parameter value k' that reduces the size of the regression tree. That is,

$$k' = \min_{t \in T_k} \bar{k}_t . \tag{18}$$

Clearly, one can re-apply this procedure to $T_{k'}$ to determine the smallest complexity parameter value k'' (and the associated tree $T_{k''}$) that will cause $T_{k''}$ to be smaller than $T_{k'}$, and so on. Therefore, starting with T_k one can repeatedly apply this procedure to find all optimally terminated trees associated with complexity parameter values larger than k. Clearly, there are, at most, $|T_k|$ such trees. This entire series of trees can be obtained from T_k without re-applying the partitioning procedure and thus can be computed quite quickly. In particular, if one uses the partitioning procedure to obtain the regression tree for k=0, all optimally terminated trees for all possible complexity parameter values can be obtained with little additional effort.

Consider the collection of all such optimally terminated trees. As $|T_k|$ becomes larger $\hat{Q}(T_k)$ becomes smaller. A plot of $\hat{Q}(T_k)$ versus $|T_k|$ usually resembles that represented in Figure 5a.

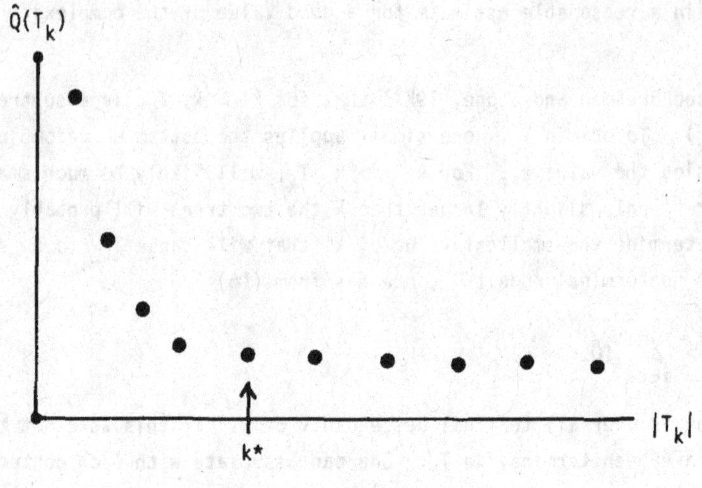

Figure 5a

There is usually a rapid decrease in the empirical residual sum of squares for the first few splits, followed by a very slow decrease with successive splits. The true residual sum of squares $Q(T_k)$ from the model tends also to decrease rapidly for the first few splits, followed by a slower decrease reaching a minimum, and then slightly increasing for even further splitting. This is illustrated in Figure 5b.

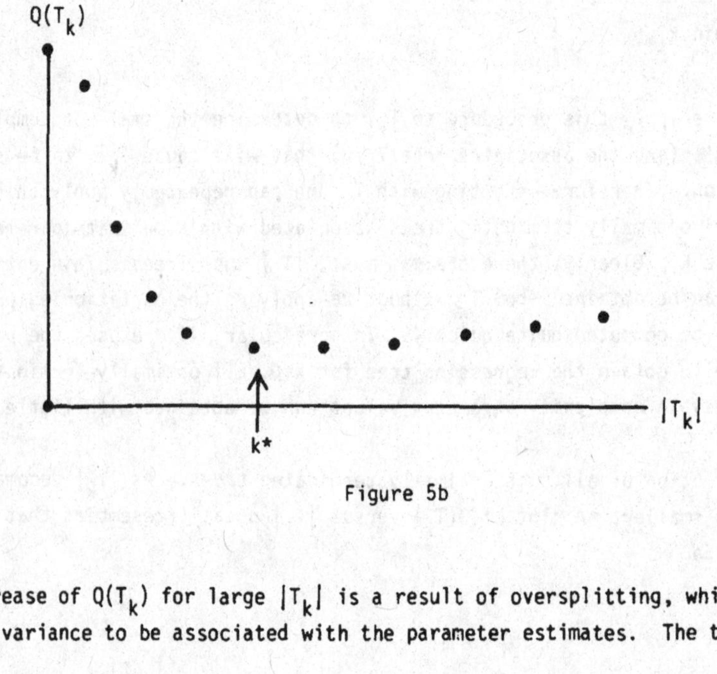

Figure 5b

The increase of $Q(T_k)$ for large $|T_k|$ is a result of oversplitting, which causes increased variance to be associated with the parameter estimates. The tree T_{k*} associated with the value k* that minimizes $Q(T_k)$ is the desired regression tree. Comparing Figures 5a and 5b, one sees that a reasonable estimate of k* can be obtained from $\hat{Q}(T_k)$ versus $|T_k|$ by choosing that value at which the decrease in $\hat{Q}(T_k)$ for increased $|T_k|$ fails to be substantial as judged from earlier decreases. Since $Q(T_k)$ versus $|T_k|$ is highly asymmetric about $|T_{k*}|$, it is wise to choose a value slightly beyond this point since slight oversplitting is much less damaging (in terms of true residual sum of squares) than undersplitting. Since $Q(T_{k*})$ is at a minimum value, values of k reasonably close to k* will cause $Q(T_k)$ to differ little from $Q(T_{k*})$. Thus, the precise value obtained for the estimate is not crucial. A good estimate for the optimum complexity parameter can thus be obtained by simple inspection of a plot of $\hat{Q}(T_k)$ versus $|T_k|$ for the collection of optimally terminated trees T_k.

The Model

The model $L_T(\underline{X})$, represented by a binary regression tree T, can be represented as

$$L_T(\underline{X}) = \sum_{t' \in T} L_{t'}(\underline{X}) \; 1(\underline{X} \in R_{t'}) \tag{19}$$

with the sum over all terminal nodes $t' \in T$. The submodel $L_{t'}(\underline{X})$ associated with each terminal node t' is linear, having the form

$$L_{t'}(\underline{X}) = \sum_{j=1}^{p} A_{t'}(j)\, X(j) + B_{t'}. \tag{20}$$

Although the parameters $A_{t'}$ and $B_{t'}$ appear linearly in (19), the global model is far from linear (unless the tree has only one node) since the regions $R_{t'}$ are determined adaptively from the training data.

By construction, the regions $R_{t'}$ associated with the terminal nodes are mutually exclusive so that for any set of carrier values $\underline{X}$ there is only one non-zero term in summation (19). Owing to the binary tree representation of the model, it is possible to determine which term will be non-zero for a given $\underline{X}$ without explicitly evaluating all of the terms in the summation. At each nonterminal node t of the tree, the split coordinate J_t and the split point s_t are stored. For any set of carrier values $\underline{X}$, the tree can be simply searched to find its corresponding terminal region. At each nonterminal node visited (starting with the root), $X(J_t)$ is compared to s_t to determine which son node to next visit:

If $X(J_t) \le s_t$: visit left son

Otherwise: visit right son.

The region $R_{t'}$ associated with the first <u>terminal</u> node t' so visited is the one containing $\underline{X}$, and the value of its associated model $L_{t'}(\underline{X})$ is the estimated response of the global model [the non-zero term in (19)]. This search procedure is illustrated in Figure 6.

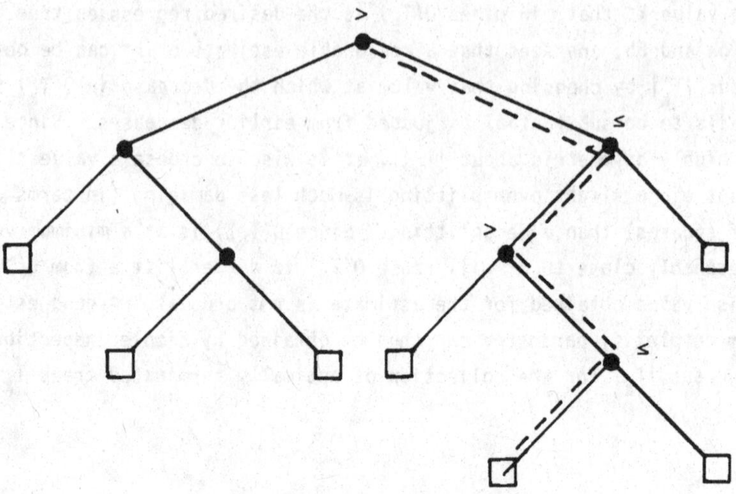

Figure 6

Speculative Splitting

The procedure described above for splitting each node, forming its two son nodes, is greedy in the sense that it tries to find that partitioning that maximizes the immediate improvement of the fit of the model to the training data represented by the node. This would be the best partitioning under the assumption that the son nodes are to be terminal and there will be no further splitting. However, for most nonterminal nodes, recursive application of the procedure accomplishes considerable further splitting.

Ideally, one would like to find the optimum sequence of cuts for improving the model fit. It is not always the case that the split that yields the best immediate improvement is the first in the best sequence of splits. Greedy strategies usually produce good solutions to optimization problems but seldom produce the optimum solutions. Finding the optimum regression tree is equivalent to binary tree optimization which is known to be NP-complete, thus requiring super polonomial computation time.

In the regression tree context, this situation arises when there are interaction effects between carriers. For example, if there is a strong interaction between carriers I and J, a single split on the Ith or Jth carrier will not significantly improve the model fit, but a split on I followed by one on J (or vise versa) will result in substantial improvement. However, a totally greedy strategy will probably fail to make that first cut (on I or J), preferring instead to cut on another carrier that yields more immediate improvement. Ultimately, the interaction will be detected by the procedure unless both the interactions and carrier designs are completely symmetric. However, the power of the procedure will be enhanced if these situations can be detected and the proper sequence of splits is made immediately.

To this end, we augment the splitting procedure at each nonterminal node, described earlier, by the following procedure. For each coordinate j, provisionally divide the subsample represented by the node at the median of X(j). The two subsamples so created are each independently fit to a complete p-variate linear model. The empirical residual sum of squares resulting from this independent piecewise linear fit is then minimized over all coordinates and the result $\hat{Q}(2p+2)$ is compared to that obtained by the univariate continuous piecewise linear procedure $\hat{Q}(3)$, described earlier. If the optimum coordinate is the same for both procedures or if

$$\hat{Q}(3) \le \frac{\#(S_t) - 3}{\#(S_t) - 2(p+1)}\ \hat{Q}(2p+2), \qquad \#(S_t) > 2(p+1) \tag{21}$$

then the split is made and the model updated in the usual manner, as described earlier. Here $\#(S_t)$ is the cardinality of the subsample represented by the node. If these conditions are both not met, then the following splitting strategy is employed. The coordinate that yielded the minimum $\hat{Q}(2p+2)$ is chosen as the split coordinate for the node. The split point is determined by provisionally splitting this coordinate at several (~10) equally spaced quantiles and finding the point that yields the best independent p-variate piecewise linear fits. However, the model augmentation parameters a_l, a_r, and b (9) are all set to zero so that there is no change in the model and, thus, no improvement in the model fit as a result of this split. This split is thus purely speculative in that by itself it results in no model improvement, but it should help to define good subsequent splits on each of its sons.

Example

In order to gain insight into the application of the partitioning procedure and the resulting regression tree model, we apply it to a computer generated training sample. Artificial rather than actual data is used so that the resulting regression tree model can be evaluated in the light of the known true underlying model. The training sample was created by generating 200 random carrier points $\{\underline{X}_i\}$ $(1 \le i \le 200)$ in the unit six-dimensional hypercube, $\underline{X}_i \in (0,1)^6$. Associated with each such vector valued carrier was a response value Y_i evaluated as

$$Y_i = 10 \sin [\pi X_i(1)\, X_i(2)] + 20\, [X_i(3) - 1/2]^2 + 10\, X_i(4) + 5\, X_i(5) + 0\, X_i(6) + \epsilon_i\ . \tag{22}$$

The set $\{\epsilon_i\}$ $(1 \le i \le 200)$ were generated as iid standard normal. For this example, the response has no dependence on one of the carriers [X(6)], a purely linear dependence on two others [X(4) and X(5)], an additive quadratic dependence on one [X(3)], and a nonlinear interaction dependence on two more [X(1) and X(2)].

The results of applying the regression tree analysis to this training sample, $\{Y_i, \underline{X}_i\}$ $(1 \le i \le 200)$, are summarized in Figures 7 and 8. The average response value is 14.3 with variance 27.1. The true mean squared error (MSE) of the best global linear least squares fit is $\sigma^2 = 7.25$, while for the regression tree this value is $\sigma^2 = 2.35$. The true intrinsic variance resulting from the noise term (ϵ) is, of course, $\sigma_I^2 = 1.0$. Figure 7 plots both the empirical MSE (from the training sample itself, solid circles) and the true MSE (open squares) as a function of tree size $|T_k|$, for all of the optimally terminated trees T_k. The value of the complexity parameter k associated with each tree is indicated above its corresponding solid circle. Inspection of Figure 7 shows, for example, that a complexity parameter value of k=0 yields a tree with 30 terminal nodes, an empirical MSE of 0.8, and a true MSE of 2.4. A value of k = 20, on the other hand, yields a tree of 10 terminal nodes, with empirical MSE 1.7, and a true MSE of 2.6. A value of $k \ge 433$ causes the tree to degenerate to solely the root node and the corresponding model is then just the global linear least squares fit. The general behavior of both the empirical and true MSE's as a function of $|T_k|$ is seen to generally correspond to that depicted in Figures 5a and 5b.

By inspecting the plot of the empirical MSE's versus $|T_k|$ (open circles) before calculating the corresponding true values (open squares), the 14 terminal node tree corresponding to k = 14 was chosen as an estimate for the optimum tree. After calculating the true MSE's, one sees that the best tree would have been the 12 terminal node tree associated with k = 16. However, any choice in the range $8 \le |T_k| \le 17$ is seen to be nearly as good from the point of view of true MSE.

Figure 8 depicts the regression tree associated with our choice of k = 14. Above the tree are shown the coefficients associated with the respective carriers [X(1) through X(6)] and the constant term, for the global linear least squares fit. This is the model L_{t_0} associated with the root node. Above each node t is the empirical residual sum of squares $\hat{Q}_t$ associated with it. Below each nonterminal node (solid circles) are shown its split coordinate J_t and split point s_t. Below each terminal node (open squares) are shown the coefficients of the first three carriers and the constant term for the model L_t associated with that node, as well as the number of training observations (circled) S_t. The values of the coefficients for the last three carriers are the same for the models associated with all nodes of the tree, as given by the global linear least squares fit.

Inspection of the binary regression tree (Figure 8) shows that the partitioning procedure behaved reasonably. It made no splits on coordinates five and six and one split on coordinate four. It made no change to the coefficients associated with these carriers from that given by the global linear least squares fit. It made substantial changes to the coefficients associated with the (first) three carriers for which the response has a highly nonlinear dependence. The first split deals with the additive nonlinear dependence by splitting the third coordinate near its central value and

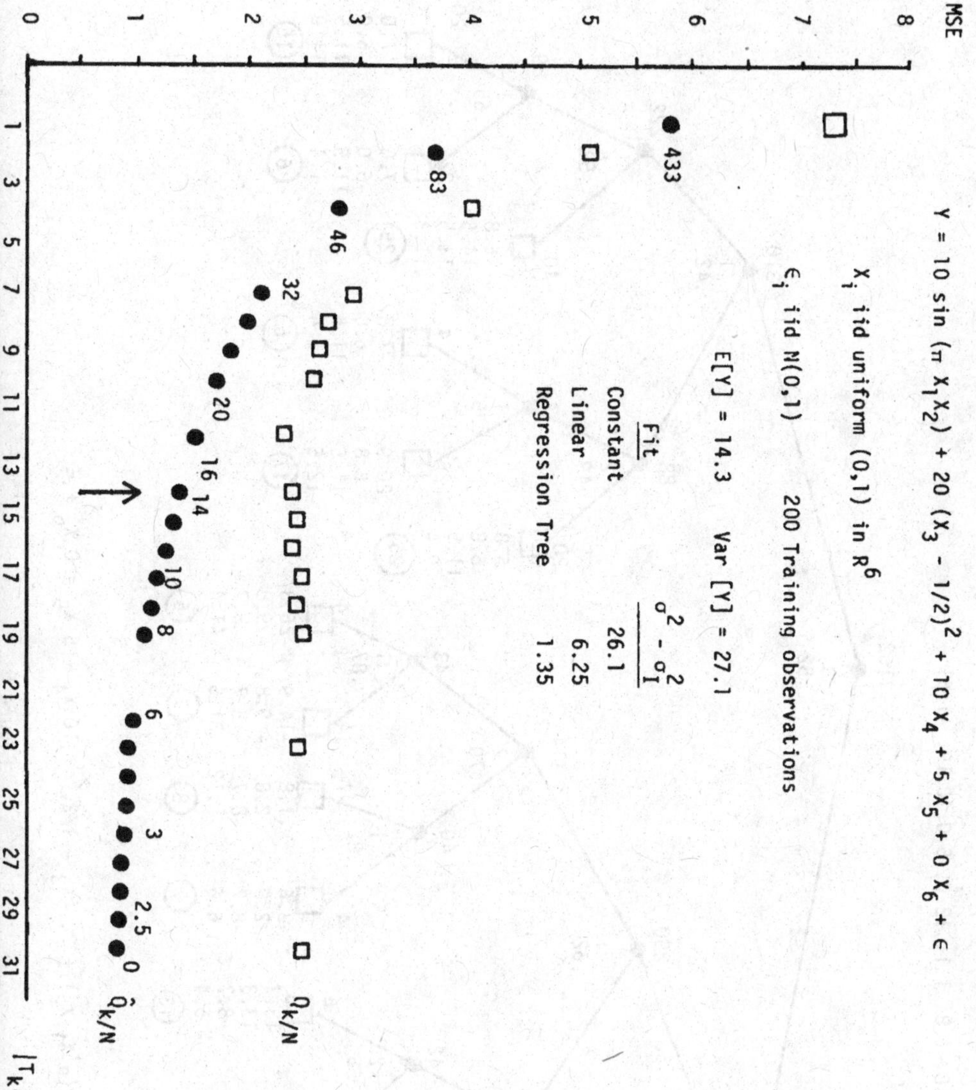

MSE
Y = 10 sin (π X1X2) + 20 (X3 - 1/2)2 + 10 X4 + 5 X5 + 0 X6 + ∈
Xi iid uniform (0,1) in R6
∈i iid N(0,1)   200 Training observations
E[Y] = 14.3   Var [Y] = 27.1
Fit
σ2 - σI2
Constant 26.1
Linear 6.25
Regression Tree 1.35
433
83
46
32
20
16
14
10
8
6
3
2.5
0
Qk/N
Q̂k/N
|Tk|

Figure 7

$L_{t_0} = (6.8, 6.8, 1.1, 12.0, 5.0, 0.19, -1.6)$

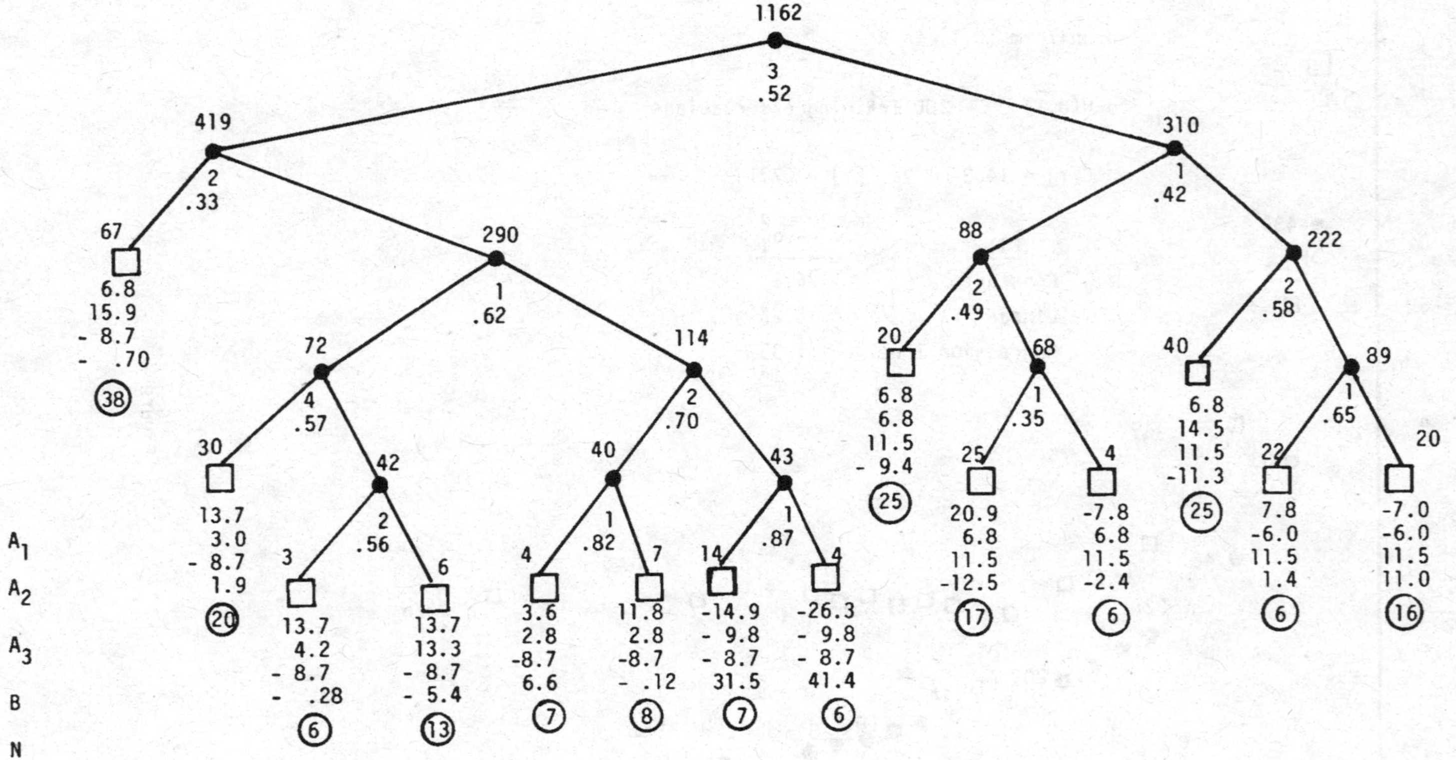

$$Y = 10 \sin(\pi X_1 X_2) + 20 (X_3 - 1/2)^2 + 10 X_4 + 5 X_5 = 0 X_6 + \epsilon$$

Figure 8

augmenting the corresponding coefficient with roughly equal and opposite slopes on each side. The remaining splits tend to alternate between the first two carriers trying to deal with their interaction. The procedure made three speculative splits ($\hat{Q}_t = \hat{Q}_{l(t)} + \hat{Q}_{r(t)}$) which tended to be well rewarded with later splits. There was only one noise split (on carrier four) for which the true underlying model had a purely linear dependence. As shown in Figure 7, the resulting regression tree model provides a much more accurate description of the training sample (in terms of true MSE) than a simple linear least squares fit.

Discussion

The example of the previous section illustrates the advantages and, to some extent, the disadvantages of the regression tree approach to nonparametric multiple regression. The basic strength of the procedure is that it is practical and seems to work. From the computational point of view, the procedure is quite fast. The time required to construct the model grows with the number of carriers p and training set cardinality N as p N log N. The time required to estimate the response value for a test observation $\underline{X}$ grows as log $|T|$ independently of p.

The statistical strength of this procedure lies with its property of being locally adaptive. It tries to partition the carrier data space into convex regions that are as large as possible subject to the constraint that a (different) linear model is a reasonable approximation to the response dependence within the region. It treats the problem (locally) in the subspace of smallest possible dimension; that is, the subspace associated with the carriers for which the response has a nonlinear dependence. In the example above, after the first cut, the procedure treated the problem mainly in the subspace of the first two carriers even through globally the problem is six-dimensional. This tends to reduce the bias associated with lack of model fit by making the most effective use of each splitting possibility. On the other hand, variance is reduced by estimating each coefficient with the largest possible training subsample. Each coefficient is estimated using all of the data associated with the largest subtree in which its corresponding carrier does not appear as a split coordinate. In the example, the entire training data set was used to estimate the coefficients of carriers 4, 5 and 6. The two coefficients associated with carrier three were each estimated using approximately one-half of the sample. The several coefficients each associated with the first two carriers are estimated using correspondingly smaller subsamples. In this way, the procedure tries to make a good trade-off between the conflicting goals of reducing both bias and variance of the model.

From the data analytic point of view, the regression tree can be interpreted as representing the nonlinear aspects of the model. The linear aspects are represented by the global linear least squares fit associated with the root node. If a global linear model adequately describes the data, then the tree will tend to collapse to only the root node. Thus, the procedure provides a loose goodness-of-fit for linear models.

Carriers that appear as split coordinates and significantly augment the model are ones for which the response dependence is highly nonlinear. (However, the converse is not necessarily true. A carrier that is highly correleated with another for which there is a highly nonlinear response dependence, will also be one for which the response has a nonlinear dependence. This carrier may never appear as a split coordinate because the splitting procedure may always prefer the carrier to which it is highly correlated). By inspecting the details of the regression tree, these inferences can be made locally in each region of the carrier data space.

Possible limitations of this regression tree approach center around its lack of continuity and robustness. The resulting model is not strictly continuous at every terminal cell boundary. It is strictly continuous at "brother" cell boundaries (those with a common parent node) but not at "cousin" cell boundaries (those for which the common ancestor is once or several times removed). However, the model is still relatively smooth at these boundaries (especially for close cousins) since the models associated with these terminal nodes share all but a few common coefficients. Still, the regression tree approach will not be appropriate in those situations for which an absolutely continuous approximation is required.

The lack of robustness associated with the procedure follows directly from its use of least squares fitting. This is easily overcome (at the expense of some computation) by simply substituting the robust/resistant analogs for the least squares routines. It is interesting to note that extreme outliers do not cause the regression tree procedure to break down (as is the case for linear least squares regression) even when least squares fitting is used. The splitting procedure tends to isolate outliers into unique terminal nodes and then proceeds with the rest of the data. Thus, although extreme outliers can seriously weaken the procedure by wasting cuts, they do not cause it to totally break down.

References

Breiman, L. and Stone, C.J. (1977). Parsimonious Binary Classification Trees. Technology Service Corporation, Santa Monica, Ca., Technical Report.

Efron, B. (1977). Bootstrap Methods: Another Look at the Jackknife. Stanford University Statistics Dept., Technical Report No. 32.

KERNEL ESTIMATION OF REGRESSION FUNCTIONS

Theo Gasser *

Zentralinstitut für Seelische Gesundheit

Abteilung Biostatistik

Postfach 5970

6800 Mannheim 1

Hans-Georg Müller **

Universität Heidelberg

Institut für Angewandte Mathematik

Im Neuenheimer Feld 294

6900 Heidelberg 1

* Research undertaken within project B 1 of the Sonderforschungsbereich 123 (Stochastic Mathematical Models) financed by the Deutsche Forschungsgemeinschaft.

** Preliminary results have been obtained in the second author's diploma at the University of Heidelberg, autumn 1978.

Summary

For the nonparametric estimation of regression functions with a one-dimensional design parameter, a new kernel estimate is defined and shown to be superior to the one introduced by Priestley and Chao (1972). The results are not restricted to positive kernels, but extend to classes of kernels satisfying certain moment conditions. An asymptotically valid solution for the boundary problem, arising for non-circular models, is found, and this allows the derivation of the asymptotic integrated mean square error. As a special case we obtain the same rates of convergence as for splines. For two optimality criteria (minimum variance, minimum mean square error) higher order kernels are explicitly tabulated.

Key words: nonparametric regression, kernel estimation, curve smoothing

1. Introduction

The nonparametric estimation of regression functions is an important data-analytic tool,for example, when data are gathered over time or space. The practical experience of the first author is based on the analysis of the Zürich longitudinal growth study (Largo et al. 1979) and on EEG-analysis (Gasser, 1977). A parametric approach is very often based on a clever guess, and not on any a priori knowledge in the field of application. Due to heterogeneity of the sample - which may be unavoidable - a single model may not be adequate. Problems of bias tend to become qualitative: an inadequate model in the study of somatic growt has for example led to the evidently erroneous conclusion, that boys and girls differ most in their prepubertal growth component (Bock et al. 1973). Nonparametric estimates show a bias, which depends in a simple, interpretative way on the function to be estimated. Nonparametric regression is usually the first, and not the only step when analysing data (norm curves in biomedicine and psychometry are exceptions to this rule). It can be used in an exploratory way to gain understanding, to arrive at an adequate parametric model, or just to substitute missing values. It yields descriptive parameters, which have the advantage of being interpretable.

The starting point of our research was a paper by Rosenblatt (1971), and the before-mentioned experience in applying smoothing methods. The related work of Benedetti (1977) and Priestley and Chao (1972) has come to our attention relatively late. We propose two new definitions for kernel estimates of regression functions; one of them is asymptotically superior to the definition introduced by Priestley and Chao (1972). Two aspects deserve our particular attention:

(i) The incorporation of the effect of the degree of non-equidistancy.

(ii) The asymptotics for kernels of increasing order, where the customary positive symmetric kernels represent the class of lowest order.

There are a number of open problems (and they will therefore be treated in some detail):

(A) Boundary effects, i.e. effects which arise for non-circular models and a finite extension of the design. They are primarily, but not exclusively, a matter of bias.

(B) Related to (A) is the search for an expression of the integrated mean square error, which contains in particular the rate of convergence.

(C) Whereas the choice of positive symmetric kernels can be guided by intuition, if is not clear which shape we should choose for higher order kernels.

Smoothing splines have gained increasing popularity among theoreticians and practicioners (among the latter: cubic splines). Kernel estimates offer an alternative which is in many ways simpler and asymptotically as efficient.

2. Model, Definitions and Criteria

Measurements $X(t_i^{(n)})$ are obtained in a sequence $t_1^{(n)}, \ldots, t_n^{(n)}$ $(0 \leq t_1^{(n)} \leq \ldots \leq t_n^{(n)} \leq 1)$ (the "design").

Let us assume that the classical regression model is valid:

$$X(t_i^{(n)}) = \mu(t_i^{(n)}) + \varepsilon_i \quad (i = 1, \ldots, n)$$

The random variables ε_i are i.i.d. with expectation zero and variance σ^2. The function $\mu(t)$ has to fulfill certain smoothness requirements to be specified below. The restriction to $\mathbb{R}^1$ is for a number of our theorems and statements easy to remove; it gives, however, more intuitive insight and is for us the most important case in application. We define and will discuss three types of kernel estimates for the estimation of $\mu(t)$:

Definition type 1: (Priestley & Chao, 1972; Benedetti, 1977)

$$(1)\quad \hat{\mu}_{n,1}(t) = \sum_{i=1}^{n} w\left(\frac{t-t_i^{(n)}}{b(n)}\right)\left(\frac{t_i^{(n)}-t_{i-1}^{(n)}}{b(n)}\right) X(t_i^{(n)})$$

Definition type 2:

$$(2)\quad \hat{\mu}_{n,2}(t) = \frac{\sum_{i=1}^{n} w\left(\frac{t-t_i^{(n)}}{b(n)}\right) X(t_i^{(n)})}{\sum_{i=1}^{n} w\left(\frac{t-t_i^{(n)}}{b(n)}\right)}$$

Definition type 3:

$$(3)\quad \hat{\mu}_{n,3}(t) = \frac{1}{b(n)} \sum_{j=1}^{n} \int_{s_{j-1}^{(n)}}^{s_j^{(n)}} w\left(\frac{t-s}{b(n)}\right) ds \cdot X(t_j^{(n)})$$

with $(s_j^{(n)})_{j=0\ldots n}$ a sequence defined as follows:

$$(4)\quad s_0^{(n)} = 0\,,\quad s_{j-1}^{(n)} \leq t_j^{(n)} \leq s_j^{(n)} \quad (j = 1,\ldots,n)\,,\quad s_n^{(n)} = 1$$

A natural choice for $s_j^{(n)}$ $(j = 1,\ldots,n-1)$ is:

$$s_j^{(n)} = \frac{1}{2}\left(t_j^{(n)} + t_{j+1}^{(n)}\right)$$

In the derivations outlined in § 3 the sequence $(s_j^{(n)})$ will also be used for definition 1. The asymptotic arguments are based on $b(n) \to 0$, $nb(n) \to \infty$ (as $n \to \infty$) throughout this paper, where $b(n)$ is a sequence of real positive bandwidths. For the kernel w ("weight function") we requir

(5) w has compact support $[-\tau, \tau]$

(6) $\int w(x)dx = 1$

(7) w fulfills a Lipschitz condition of order γ_w $(0 < \gamma_w \leq 1)$

There are a number of reasons for restricting kernels to compact support

(i) This allows simpler proofs with more powerful asymptotic results.

(ii) Optimal kernels and minimum variance kernels (§ 5) have compact support.

(iii) Boundary effects without such an assumption are smeared over the whole interval [0,1] (compare § 6).

(iv) If the kernel has to be coded, we must truncate a kernel with non-compact support.

For an equidistant design, there is not much to choose between definitio 1-3. It turns out, that the asymptotic properties of type 3 are superior to the others in the general case; one would also guess that type 3 is superior to type 1 in a finite sample situation. The type 2 definition suffers from the drawback that the denominator may become zero, if negative weights are allowed. In the rest of the paper, we will concentrate on the type 3 estimate, and we will then omit the subscript. Note that the integral needed in definition 3 is known analytically for the kernels of practical interest. The definition given is appropriate for $t \in [b(n), 1-b(n)]$ (the "interior" of the smoothing method); the estimation for $t \in [0,b(n))$, or $t \in (1-b(n),1]$ (the "boundary") will be discussed in § 6.

The basic requirement for the design is:

$$(8) \qquad \max_j \left| t_j^{(n)} - t_{j-1}^{(n)} \right| = O\left(\frac{1}{n}\right)$$

but at some points we will require a form of asymptotic equidistance with rate $\delta > 1$:

$$(9) \qquad \max_j \left| s_j^{(n)} - s_{j-1}^{(n)} - \frac{1}{n} \right| = O\left(\frac{1}{n^\delta}\right)$$

For an equidistant design we can put the terms involving $O\left(\frac{1}{n^\delta}\right)$ equal to zero. The criterion we use is the mean square error (MSE) evaluated at a point:

$$E(\hat{\mu}_n(t) - \mu(t))^2$$

or the integrated mean square error (IMSE) as a global measure:

$$\int_0^1 E(\hat{\mu}_n(t) - \mu(t))^2 \, dt$$

3. Local Asymptotic Results

As a first step, we give an integral approximation for $E(\hat{\mu}_{n,i}(t))$ $(i = 1,2,3)$, assuming μ to be Lipschitz continuous of order γ_μ:

$$(10) \quad E(\hat{\mu}_{n,i}(t)) = \frac{1}{b(n)} \int_0^1 w\left(\frac{t-s}{b(n)}\right) \mu(s)ds + O(f(n,b(n)),\gamma_\mu,\gamma_w)$$

The derivation, outlined in appendix 1, leads to the following approximation errors:

$$(10)' \quad O(f(n,b(n),\gamma_\mu,\gamma_w) = O\left(\frac{1}{n^{\gamma_\mu}} + \frac{1}{n^{\gamma_w} b(n)^{\gamma_w}}\right), \text{ estimate type 1}$$

$$= O\left(\frac{1}{n^{\gamma_\mu}} + \frac{1}{n^{\gamma_w} b(n)^{\gamma_w}} + \frac{1}{n^{\delta-1}}\right) \quad \text{" type 2}$$

$$= O\left(\frac{1}{n^{\gamma_\mu}}\right) \quad , \text{ estimate type 3}$$

The newly proposed type 3 estimate has a better remainder term for the bias than the comparable estimate 1; since there is no asymptotic difference with respect to the variance, the type 3 estimate is strictly superior to the definition 1 (Priestley and Chao, 1972). The somewhat different type 2 estimate needs a relatively strict form of asymptotic independence to yield a tolerable remainder term. We note without proof that the remainder terms have to be multiplied by $\frac{1}{b(n)}$ when we drop the assumption of compact support.

The integral approximation for the variance is the same for the type 1 and the type 3 estimates (appendix 2):

$$(11)\quad \mathrm{Var}(\hat{\mu}_{n,i}(t)) = \frac{\sigma^2}{nb(n)} \int_{-\tau}^{\tau} w^2(x)dx + O\left(\frac{1}{n^{1+\gamma_w} b(n)^{1+\gamma_w}} + \frac{1}{n^{\delta} b(n)}\right)$$

$$i = 1,3$$

For the type 2 estimate, the second remainder term becomes $\frac{1}{n^{\delta-1}}$.

In the following we restrict ourselves to the type 3 definition.

Definition: A kernel w satisfying (5)-(7) is called a kernel of order k, if the following holds:

$$(12)\quad \int_{-\tau}^{\tau} w(x)x^j\, dx = 0 \qquad j = 1,\dots,k-1$$

$$(13)\quad \int_{-\tau}^{\tau} w(x)x^k\, dx = \beta_k \neq 0$$

In what follows, the kernel w is assumed to be Lipschitz-continuous of order $\gamma_w = 1$ (with the exception of a finite number of points at most).

Theorem 1 (Bias)

Let us assume that w is a kernel of order k, and that the regression function $\mu(t)$ is m times differentiable with a continuous m-th derivative on $[0,1]$ $(m \geq k)$. Then the bias for all $t \in (0,1)$ can be expressed as follows:

$$(14)\quad E(\hat{\mu}_n(t) - \mu(t)) = \frac{(-1)^k}{k!} b(n)^k \int_{-\tau}^{\tau} x^k w(x)dx\, \mu^{(k)}(t)$$

$$+ O\left(\frac{1}{n}\right) + o(b(n)^k)$$

If $m > k$, we obtain $O(b(n)^{k+1})$ for the second term.

Proof: Based on the integral approximation (10) we obtain:

$$E(\hat{\mu}_n(t) - \mu(t)) = \int_{(t-1)/b(n)}^{t/b(n)} w(x)\mu(t-b(n)x)dx - \mu(t) + O\left(\frac{1}{n}\right)$$

Due to the compactness of support, we have $\left[\frac{t-1}{b(n)}, \frac{t}{b(n)}\right]$ $\supset [-\tau,+\tau]$ for some $n_o > n$, and hence:

$$= \int_{-\tau}^{\tau} w(x)(\mu(t-b(n)x) - \mu(t))dx + O\left(\frac{1}{n}\right)$$

Taylor's formula and the orthogonality conditions of w yield:

$$(0 < \theta < 1) = \frac{(-1)^k}{k!} b(n)^k \int_{-\tau}^{\tau} w(x)\mu^{(k)}(t-\theta b(n)x)x^k dx + O\left(\frac{1}{n}\right)$$

From this follows the proposition by the continuity of $\mu^{(k)}$.

Remarks:

1. For a polynomial μ of order (k-1) or less, the bias reduces to the approximation error. With an equidistant design, and if (12),(13) are valid for discrete moments, the bias for such polynomials is finitely zero.

2. Those kernels used today are symmetric and positive, and are therefore of order $k = 2$.

3. It is often argued that the shape of a kernel is unimportant, whereas the choice of the smoothing parameter b is most important. The first statement is correct only within a class of kernels of order k.

<u>Theorem 2</u> (Mean square error, consistency)

If the assumptions of theorem 1 are valid, and if:

$$\max_j \left| s_j^{(n)} - s_{j-1}^{(n)} - \frac{1}{n} \right| = O\left(\frac{1}{n^\delta}\right) \quad , \delta > 1$$

we have for all $t \in (0,1)$ for the mean square error:

$$(15) \quad E(\hat{\mu}_n(t) - \mu(t))^2 = \frac{\sigma^2}{nb(n)} \int_{-\tau}^{\tau} w(x)^2 dx +$$

$$\frac{b(n)^{2k}}{k!^2} \left(\int_{-\tau}^{\tau} w(x) x^k dx \right)^2 \mu^{(k)}(t)^2 + O\left(\frac{1}{n^\delta b(n)} + \frac{1}{n^2 b(n)^2}\right) + o\left(\frac{b(n)^k}{n} + (b(n)^{2k}\right)$$

If $m > k$, the last term becomes $O(b(n)^{2k+1} + \frac{b(n)^{k+1}}{n})$

Consistency in MSE is guaranteed if $nb(n) \to \infty$, $b(n) \to 0$ $(n \to \infty)$

Proof requires theorem 1, and relation (11).

Remarks:

1. If one is interested only in consistency, the requirements on μ and w can be relaxed to Lipschitz-continuity of order γ_μ and γ_w, and the assumption (8) is sufficient.
2. Consistency has been previously proved under less favourable conditions (Priestley and Chao, 1972, assume $nb(n)^2 \to \infty$), and this is due to the non-compact support.

Corollary: The asymptotically optimal smoothing parameter $b^*(n)$ with respect to MSE is as follows:

$$(16) \quad b^*(n) = \left(\frac{1}{2k} \cdot \frac{k!^2 \sigma^2 \int_{-\tau}^{\tau} w(x)^2 dx}{\left(\int_{-\tau}^{\tau} w(x) x^k dx\right)^2 \mu^{(k)}(t)^2} \cdot \frac{1}{n}\right)^{\frac{1}{2k+1}}$$

where $\mu^{(k)}(t) \neq 0$

The MSE at the optimal bandwidth becomes:

$$(17) \quad n^{-\frac{2k}{2k+1}} c(k,\sigma^2) \left[\mu^{(k)}(t) \left[\int_{-\tau}^{\tau} w(x)^2 dx\right]^k \int_{-\tau}^{\tau} w(x) x^k dx\right]^{\frac{2}{2k+1}}$$

$$+ o\left(n^{-\frac{2k}{2k+1}}\right)$$

Proof: We have to take the first derivative with respect to b(n) of the leading terms in (15) to obtain (16).

Remark:

The kernel estimates have asymptotic efficiency zero relative to parametric regression, when the model is known. A more interesting question is, what the relations will be, when only some assumptions about the smoothness of μ can be made. Formula (17) shows that the MSE improves with increasing kernel order for a large enough sample size; a Monte Carlo study will give information for which n, and μ, this gain materializes.

Benedetti (1977) has proved almost sure convergence and asymptotic normality for the type 1 estimate with some restrictions on the kernels. These results extend to the type 3 estimate, but only almost sure convergence will be treated.

Theorem 3 (Almost sure convergence)

Assume Lipschitz-continuity of order γ_μ, γ_w for μ and w.

If $E(\varepsilon_i^4) < \infty$ and $\sum_{n=1}^{\infty} \frac{1}{n^2 b(n)^2} < \infty$, we have $\hat{\mu}_n(t) \to \mu(t)$ almost surely as $n \to \infty$.

Proof: Let us apply the Čebysev inequality, using the integral approximation for the variance

$$\sum_{n=1}^{\infty} P(|\hat{\mu}_n(t) - E(\hat{\mu}_n(t))| > c) \leq \frac{E(\varepsilon_1^4)}{c^4} \sum_{n=1}^{\infty} \left[O\left(\frac{1}{nb(n)}\right)\right]^2$$

From the Borel-Cantelli lemma, we conclude $(\hat{\mu}_n(t) - E(\hat{\mu}_n(t))) \to 0$ a.s., and the assertion follows from the asymptotic unbiasedness of the estimate $\hat{\mu}_n(t)$.

Remark:

The assumption is amply fulfilled for the optimal bandwidth for kernels of arbitrary order.

4. Global Asymptotic Results

The global asymptotic behaviour of the estimate, as measured by the integrated mean square error (IMSE), is of interest for a number of reasons:

(i) To obtain kernels which have good properties globally (compare § 5)

(ii) When devising methods to choose the smoothing parameter b from the data, we should check them against a global measure of quality.

(iii) For the comparison of different methods, e.g. with respect to the rate of convergence.

Whenever we can assume $\mu(t)$ to be a periodic function, the IMSE can be easily derived from the MSE (theorem 2):

$$(18)\quad \mathrm{IMSE} = \frac{\sigma^2}{nb(n)} \int_{-\tau}^{\tau} w^2(x)dx + \frac{1}{k!^2}\, b(n)^{2k} \left(\int_{-\tau}^{\tau} w(x)x^k dx\right)^2$$

$$\int_0^1 \mu^{(k)}(t)^2\, dt + O\left(\frac{1}{n^{\delta} b(n)} + \frac{1}{n^2 b(n)^2}\right) + o\left(\frac{b(n)^k}{n} + o(b(n)^k)\right)$$

if $\mu^{(m)}$ exists for some $m > k$, the last term becomes $O(b(n)^{2k+1} + \frac{b(n)^{k+1}}{n})$

The extension of this result to a non-periodic model is a substantial problem, due to boundary effects which occur for $t \in [0,b(n)\cdot\tau)$ and for $t \in (1-b(n)\cdot\tau,1]$. For boundary values we have $[t-b(n)\cdot\tau,t+b(n)\tau] \not\subset [0,1]$. The boundary disappears asymptotically (as $b(n) \to 0, n \to \infty$), but its influence dominates the global asymptotic behaviour for estimate (3) (and also for the "cut-and-normalize" modification given in § 6). The integrated mean square error can be obtained without modification of the estimate for an interval $[\varepsilon,1-\varepsilon]$ (for some $0 < \varepsilon < 1$) but not for $[0,1]$ as in (18). This is based on theorems 1 and 2 and the following remark: there is an $\varepsilon > 0$ and a n_0 such that for all $n > n_0$, and all fixed $t \in [\varepsilon,1-\varepsilon]$ we have $[-\tau,+\tau] \subset [(t-1)/b(n),t/b(n)]$.

In order to investigate the asymptotics on the boundary interval, we consider a sequence $t(n)$ which has the same relative location in the boundary interval for all n:

$$t(n) = q\, b(n) \qquad q \in [0,1]$$
$$t(n) = 1-q\, b(n)$$

For the sequence $t(n) = qb(n)$ that part of the support of the kernel which is mapped into $[0,1]$ is $[-\tau, q\tau]$. Our solution to the boundary problem proposed in § 6 consists in the introduction of modified kernels w_q, $q \in [0,1)$, with support $[-\tau, q\tau]$ and satisfying the moment conditions of the interior. This allows us to determine the IMSE on $[0,1]$, also in the noncircular case:

<u>Theorem 4</u> (IMSE, rate of convergence)

Let us assume w to be Lipschitz continuous of order 1, and μ to be a function as in theorem 1. The kernel w_q used for the sequence $t(n)=qb(n)$ or $t(n) = 1-qb(n)$ at the boundary $(0 \leq q < 1)$, has to satisfy the conditions (i) - (iii) of § 6 . For the grid we assume:

$$\max_j \left| s_j^{(n)} - s_{j-1}^{(n)} - \frac{1}{n} \right| = O\left(\frac{1}{n^\delta}\right) \qquad \delta > 1$$

$$(18)' \quad IMSE = \frac{\sigma^2}{nb(n)} \int_{-\tau}^{\tau} w(x)^2 dx + \frac{1}{k!^2} b(n)^{2k} \left(\int_{-\tau}^{\tau} w(x)x^k dx\right)^2 \cdot$$

$$\int_0^1 \mu^{(k)}(t)^2 dt + O\left(\frac{1}{n^\delta b(n)} + \frac{1}{n^2 b(n)^2}\right) + o\left(\frac{b(n)^k}{n} + o(b(n)^k)\right)$$

whenever $\mu^{(m)}$, $m > k$, exists, the last term becomes: $O\left(b(n)^{2k+1} + \frac{b(n)^{k+1}}{n}\right)$

The asymptotically optimal smoothing parameter is:

$$(19)\quad b^*(n) = \left(\frac{k!^2\sigma^2 \int_{-\tau}^{\tau} w(x)^2\,dx}{2k\left(\int_{-\tau}^{\tau} w(x)x^k dx\right)^2 \int_0^1 \mu^{(k)}(t)^2 dt} \cdot \frac{1}{n} \right)^{\frac{1}{2k+1}}$$

The IMSE at the optimal value $b^*(n)$ is :

$$(20)\quad \text{IMSE} = n^{-\frac{2k}{2k+1}}\, c(k,\sigma)\left(\int_{-\tau}^{\tau} w(x)x^k dx \left(\int_{-\tau}^{\tau} w(x)^2 dx\right)^k \right)^{\frac{2}{2k+1}} \left(\int_0^1 \mu^{(k)}(t)^2 dt \right)^{\frac{1}{2k+1}} + o\left(n^{-\frac{2k}{2k+1}}\right)$$

if $\mu^{(m)}$, $m > k$, exists, and if $\delta > (2k+2)/(2k+1)$, one has a remainder term $O\left(\frac{1}{n}\right)$.

<u>Proof:</u> A proof for relation (18) is given and from relation (18) the relations (19),(20) can be deduced. We first note, that the integral approximations (10),(11) remain valid for a sequence of points $t(n)$. It is sufficient to make the argument for $t(n) = q\, b(n)$, the left boundary. From the requirements (i),(ii), § 6, we conclude:

$$\text{Var}(\hat{\mu}_n(t(n))) = \frac{\sigma^2}{nb(n)} \int_{-\tau}^{q\tau} w_q(x)^2 dx + O\left(\frac{1}{n^2 b(n)^2} + \frac{1}{n^\delta b(n)}\right)$$

$$(E(\hat{\mu}_n(t) - \mu(t)))^2 = \frac{b(n)^{2k}}{k!^2}\left[\int_{-\tau}^{q\tau} w_q(x)x^k dx\right]^2 \mu^{(k)}(o)^2 + o(b(n)^{2k}) + o\left(\frac{b(n)^k}{n}\right)$$

Using relation (iii), we obtain

$$\int_0^{b(n)} \mathrm{Var}(\hat{\mu}_n(t(n)))dt(n) \leq \frac{\sigma^2}{n} \max_{q\in[0,1)} \int_{-\tau}^{q\tau} w_q(x)^2 dx + O\left(\frac{1}{n^2 b(n)} + \frac{1}{n^\delta}\right)$$

$$= O\left(\frac{1}{n} + \frac{1}{n^2 b(n)} + \frac{1}{n^\delta}\right)$$

$$\int_0^{b(n)} E(\hat{\mu}_n(t(n)) - \mu(t(n))))^2 dt(n) \leq \frac{b(n)^{2k+1}}{k!^2} \max_{q\in[0,1)}{}^1 \left(\int_{-\tau}^{q\tau} w_q(x)x^k dx\right)^2 \mu^{(k)}(0$$

$$+ o\,(b(n)^{2k+1}) + o\left(\frac{b(n)^{2k+1}}{n}\right)$$

$$= o\,(b(n)^{2k+1})$$

The IMSE then becomes:

$$\mathrm{IMSE}([0,1]) = \mathrm{IMSE}([b(n),1-b(n)]) + \mathrm{IMSE}([0,b(n))) + \mathrm{IMSE}((1-b(n),1])$$

$$= \frac{\sigma^2(1-2b(n))}{n\,b(n)} \int_{-\tau}^{\tau} w(x)^2 dx + \frac{b(n)^{2k}}{k!^2} \left(\int_{-\tau}^{\tau} w(x)x^k dx\right)^2 .$$

$$\int_{b(n)}^{1-b(n)} \mu^{(k)}(t)^2 dt + O\left(\frac{1}{n^\delta b(n)} + \frac{1}{n^2 b(n)^2} + \frac{1}{n}\right) + o\,(b(n)^{2k}$$

The assertion follows by observing that

$$b(n)^{2k}\cdot\mathrm{const}\cdot \int_0^{b(n)} \mu^{(k)}(t)^2 dt = O\,(b(n)^{2k+1})$$

This proof implies that it does not matter globally what solutions we choose at the boundary, as long as their asymptotic behaviour is equivalent to that of the interior.

Remarks:

1. The existence of a class of boundary kernels satisfying (i)-(iii) follows from theorem 8.
2. The remainder term of the optimal IMSE is dominated by the approximation of the bias for $\delta > (2k + 2)/(2k + 1)$, and by the variance for $0 < \delta < (2k + 2)/(2k + 1)$. Note that the leading term and the remainder term come closer for increasing k. This is a warning not to overestimate the validity of asymptotic expressions for higher order kernels in a finite sample situation.
3. Craven and Wahba (1979) have given an upper bound for the IMSE for polynomial spline smoothing and have obtained a rate of convergence $O(n^{-\frac{2m}{2m+1}})$ for a spline of degree $(2m - 1)$.

5. Hierarchies of kernels

The foregoing chapters underline the importance of distinguishing classes of kernels to the order k. Kernels of second order used exclusively today lead to a bias proportional to the second derivative which is large at points of interest, as for example peaks in a curve. The choice of these "standard kernels" can be guided by intuition, and this is not true for higher order kernels. In the following we will require the kernels to be symmetric; the assumption of compact support is not crucial for this chapter. Kernels of order k are defined to be optimal, when they minimize the asymptotic IMSE (20), i.e. the following functional:

$$(21) \quad T(W) = \left| \int_{-\tau}^{\tau} w(x) x^k dx \right| \left(\int_{-\tau}^{\tau} w(x)^2 dx \right)^k$$

with the side conditions (12),(13)

Definition Let w_o be a kernel with support $[-1,+1]$. Then the classes of kernels $w_\tau(x) = \frac{1}{\tau} w_o\left(\frac{x}{\tau}\right)$ is called an equivalence class.

Lemma 1 The functional $T(w)$ is invariant for an equivalence class of kernels.

Proof: w_τ has compact support $[-\tau,+\tau]$.The lemma follows by substituti $\frac{x}{\tau} = s$ in the integrals.

Remark:

Without restriction of generality, we may assume support $[-1,+1]$ whenever no assumptions regarding $\beta_k = \int_{-\tau}^{\tau} w_\tau(x) x^k dx$ or $\int_{-\tau}^{\tau} w_\tau(x)^2 dx$ need be made.

Assuming symmetry and positivity, Epanechnikov (1969) derived the optim kernel for estimating probability densities:

$$w_2(x) = .75(1 - x^2) \qquad |x| \leq 1$$
$$= 0 \qquad |x| > 1$$

For regression functions (estimate type 2), Benedetti (1977) has obtaine the weaker result, that this kernel is optimal with respect to MSE. Variational calculus is the method of choice to find extremal solutions for the functional T with side conditions (12),(13):

$$(22)\quad V(w) = \int_{-\tau}^{\tau} w(x)^2 dx = \min!$$

$$(23)\quad \int_{-\tau}^{\tau} w(x)x^j dx = 1 \qquad j = 0$$

$$= 0 \qquad j = 1, \dots, k-1$$

$$= (-1)^{\frac{k}{2}+1} \qquad j = k$$

Note that by lemma 1 one of the parameters τ or $\beta_k = \int_{-\tau}^{\tau} w(x)x^k dx$ may be normalized without affecting the solution. As a solution to the extremal problem, we obtain a polynomial with properties to be specified in theorem 5.

$$(24)\quad w(x) = \sum_{i=0}^{k} \lambda_i x^i = p_k(x) \qquad |x| \leq \tau$$

$$= 0 \qquad |x| > \tau$$

The conditions (23) lead to a system of (k+1) linear equations for the (k+1) coefficients $\lambda_0, \dots, \lambda_k$:

$$(25)\quad 2\begin{bmatrix} \tau & \frac{\tau^2}{2} & \cdots & \frac{\tau^{k+1}}{k+1} \\ \frac{\tau^2}{2} & \frac{\tau^3}{3} & \cdots & \frac{\tau^{k+2}}{k+2} \\ \vdots & \vdots & & \vdots \\ \frac{\tau^{k+1}}{k+1} & \frac{\tau^{k+2}}{k+2} & \cdots & \frac{\tau^{2k+1}}{2k+1} \end{bmatrix} \begin{bmatrix} \lambda_0 \\ \lambda_1 \\ \vdots \\ \lambda_k \end{bmatrix} = \begin{bmatrix} 1 \\ 0 \\ \vdots \\ (-1)^{\frac{k}{2}+1} \end{bmatrix}$$

The matrix is regular and there exists a unique solution. The support $[-\tau,+\tau]$ is not yet fixed, and by a suitable choice of τ one can make T in (21) as small as one likes (compare e.g. minimum variance kernels). The concept of positivity is therefore generalized so as to be useful for higher order kernels:

Definition A kernel of order k is called minimal if it has a minimal number of changes of sign on its support.

Lemma 2 A symmetric kernel of order k is minimal if it has (k-2) changes of sign on its support.

Proof: The interval $[-\tau,+\tau]$ is decomposed into 2j subintervals $I_1, \ldots, I_j, I_{j+1}, \ldots, I_{2j}$ defined by the points where a change of sign occurs, and by $-\tau,+\tau$ and 0 . Due to the symmetry of the kernel, we can restrict the argument to $[-\tau, 0]$ and moments of even order. We prove by contradiction that:

$$j \geq \frac{k}{2}$$

The following matrix is regular by construction:

$$M = \begin{bmatrix} \int_{I_1} w(x)x^2dx & \cdots & \int_{I_j} w(x)x^2dx \\ \vdots & & \vdots \\ \int_{I_1} w(x)x^{2j}dx & \cdots & \int_{I_j} w(x)x^{2j}dx \end{bmatrix}$$

If M were singular, there exist $\{\alpha_1, \ldots, \alpha_j\} \neq \{0, \ldots, 0\}$ such that:

$$\int_{I_k} \sum_{i=1}^{j} \alpha_i x^{2i} w(x)dx = 0 \qquad (k = 1,\ldots,j)$$

Since w does not change its sign in I_k , the symmetric polynomial:

$$q(x) = \sum_{i=1}^{j} \alpha_i x^{2i}$$

has a root within the interval I_k . In total this gives 2j roots not equal to zero, and a root at zero; we conclude that $\{\alpha_1,\ldots,\alpha_j\} = \{0,\ldots 0\}$ If $j < \frac{k}{2}$, the matrix M would map the vector $\{1,\ldots,1\}$ into zero by the moment conditions, and this is a contradiction to the regularity of M . Regarding the existence of kernels of order k which satisfy $j = \frac{k}{2}$, we refer the reader to the minimum variance kernels introduced below.

<u>Remarks:</u>

It follows from this lemma, that a minimal polynomial kernel of order k has to be of degree $\geq (k-2)$ with at least (k-2) real roots (compare with the minimum variance kernels). If the support is defined by the outermost roots of a polynomial, it has to be of degree $\geq$ k with k real roots.

For finding optimal kernels which have a minimal number of sign changes, we tentatively replace the system (25) by the following system of linear equations (here the support is normalized to [-1,+1]):

$$(26) \qquad 2\begin{bmatrix} 1 & \frac{1}{3} & \cdots & \frac{1}{k+1} \\ \frac{1}{3} & \frac{1}{5} & \cdots & \frac{1}{k+3} \\ \vdots & \vdots & & \vdots \\ \frac{1}{k-1} & \frac{1}{k+1} & \cdots & \frac{1}{2k-1} \\ 1 & 1 & & 1 \end{bmatrix} \begin{bmatrix} \lambda_o \\ \lambda_2 \\ \vdots \\ \lambda_{k-2} \\ \lambda_k \end{bmatrix} = \begin{bmatrix} 1 \\ 0 \\ \vdots \\ 0 \\ 0 \end{bmatrix}$$

Theorem 5

A. The matrix in (26) is regular.

B. The coefficients $\lambda_i (i = 0,2, \ldots , k)$ of the polynomial p_k determined from (26) are all different from zero and alternate in sign ($sgn(\lambda_o) > 0$). The polynomial has k real roots of multiplicity one.

Remark:

From A. follows that there is exactly one polynomial kernel of order k with a root at the boundary of support $\tau = 1$, which satisfies the side conditions (12),(13) and the requirement of minimality.

Proof:

The following matrix M , which differs in the last row from the matrix in (26) , is regular:

$$M = \begin{bmatrix} 1 & \frac{1}{3} & \cdots & \frac{1}{k+1} \\ \vdots & \vdots & & \vdots \\ \frac{1}{k+1} & \frac{1}{k+3} & \cdots & \frac{1}{2k+1} \end{bmatrix}$$

The vectors $1, x^2, \dots, x^k$ are linearly independent in the space $L_2[-1,+1]$. The kernel w is by (24) a linear combination of these vectors, and the scalar product of w with $1, x^2, \dots, x^k$ has given values. As a consequence, the following system has exactly one solution for each $\beta \in \mathbb{R}$:

$$(27) \qquad 2M \begin{bmatrix} \lambda_o \\ \lambda_2 \\ \vdots \\ \lambda_k \end{bmatrix} = \begin{bmatrix} 1 \\ 0 \\ \vdots \\ \beta \end{bmatrix}$$

If $\{\lambda_o^*, \dots, \lambda_k^*\}$ is a solution of (26) it is also a solution of (27) for $\beta^* = \frac{2}{k+1} \lambda_o^* + \dots + \frac{2}{2k+1} \lambda_k^*$. It remains to be proved that there is exactly one β' such that the solution $\{\lambda_o', \dots, \lambda_k'\}$ of (27) satisfies $s = \lambda_o' + \dots + \lambda_k' = 0$. We have $s(\beta) = c_1 + c_2\beta$ since $\lambda_o, \dots, \lambda_k$ depend linearly on β. The assertion is true if $s(\beta)$ changes its sign. Choose $\beta_1 = 0$. According to lemma 2 $p_1(x) = \hat{\lambda}_o + \hat{\lambda}_2 x^2 + \dots + \hat{\lambda}_k x^k$ has exactly k changes of sign, and at each root of $p_1(x)$ there has to be a change of sign.

$$(28) \qquad \operatorname{sgn} s(\beta_1) = \operatorname{sgn} p_1(1) = (-1)^{\frac{k}{2}} \operatorname{sgn} p_1(0)$$

Consider a reduced system instead of (27):

$$(27)' \quad 2\begin{bmatrix} 1 & \frac{1}{3} & \cdots & \frac{1}{k-1} \\ \vdots & \vdots & & \vdots \\ \frac{1}{k-1} & \frac{1}{k+1} & & \frac{1}{2k-3} \end{bmatrix} \begin{bmatrix} \alpha_o \\ \vdots \\ \alpha_{k-2} \end{bmatrix} = \begin{bmatrix} 1 \\ 0 \\ \vdots \\ 0 \end{bmatrix}$$

and choose $\beta_2 = \frac{2}{k+1}\alpha_o + \ldots + \frac{2}{2k-1}\alpha_{k-2}$. With such a $\beta = \beta_2$, the solution of (27)' coincides with that of (27) : $p_2(x) = \alpha_o + \alpha_2 x^2 + \ldots + \alpha_{k-2} x^{k-2}$. Due to lemma 2 , it has (k-2) changes of sign and it is a polynomial of degree (k-2) with:

$$(28)' \quad \text{sgn } s(\beta_2) = \text{sgn } p_2(1) = (-1)^{\frac{k-2}{2}} \text{ sgn } p_2(0)$$

The proof of $\text{sgn } \alpha_o = \text{sgn } \hat{\lambda}_o = 1$ completes the proof of A:

$$0 < \int_{-1}^{+1} p_1(x)^2 dx = \int_{-1}^{+1} p_1(x)(\hat{\lambda}_o + \hat{\lambda}_2 x^2 + \ldots + \hat{\lambda}_k x^k) dx = \hat{\lambda}_o$$

and similarly $$0 < \int_{-1}^{1} p_2(x)^2 dx = \alpha_o$$

The proof of B is based on lemma 2: The polynomial p_k has to have k real roots of multiplicity one, since there are (k-2) changes of sign in the interior, and a root both at -1 and at +1 . A fortiori, the polynomial p_k is of degree k . That $\lambda_i \neq 0 (i = 0,\ldots,k)$ can be seen from the decomposition of the polynomial p_k into linear factors. From the same argument follows the alternation of the sign of the coefficients. It remains to prove $\text{sgn}(\lambda_o) > 0$. For the solution p_k of (26) we have:

$$\beta_1 \leq \beta \leq \beta_2$$

where β is the k-th moment for $w = p_k[-1,+1]$. The coefficient λ_o depends linearly on the k-th moment, and both p_1 and p_2 have a positive zero coefficient, hence $\lambda_o > 0$.

Solving the system of equations (26) for orders k = 2,4,6,8 yields the kernels given numerically in table 1, and graphically in fig. 1. The expressions relevant for the asymptotic variance, the asymptotic bias, and the asymptotic MSE/IMSE are included in table 1:

$$V = \int_{-1}^{1} w(x)^2 dx$$

$$B = \int_{-1}^{1} w(x) x^k dx$$

$$T = (B \cdot V^k)^{\frac{2}{2k+1}}$$

The simplicity of the analytical form of V and B is notable:

(29) $$V = (\lambda_o + \lambda_k B)$$

(30) $$B = 2 \sum_{i=o}^{k} \lambda_i / (i + k + 1)$$

Order k	w(x)	B	V	T
2	$0.75 - 0.75\, x^2$	+.20000	0.600	0.3491
4	$1.406 - 4.687\, x^2 + 3.281\, x^4$	-.04762	1.250	0.6199
6	$2.051 - 14.35\, x^2 + 25.83\, x^4 - 13.53\, x^6$	+.01166	1.893	0.9086
8	$2.690 - 32.26\, x^2 + 106.4\, x^4 - 131.6\, x^6 + 54.81\, x^8$	-.00287	2.533	1.204

Table 1: Optimal kernels of order 2-8 on [-1,+1]

Table 2: Minimum variance kernels of order 2-8 on [-1,+1]

Order k	w(x)	B	V	T
2	0.5	+.33333	0.500	0.3701
4	$1.125 - 1.875\, x^2$	-.08571	1.125	0.6432
6	$1.758 - 8.203\, x^2 + 7.383\, x^4$	+.02164	1.758	0.9333
8	$2.392 - 21.53\, x^2 + 47.37\, x^4 - 29.32\, x^6$	-.00543	2.392	1.231

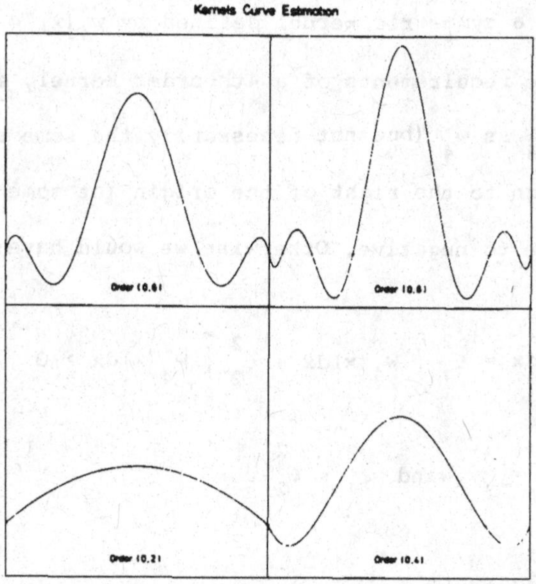

fig. 1: Optimal kernels of order 2-8, common min-max-scale, dotted line at zero

We note that for k = 2 we obtain the known optimal positive kernel (Epanechnikov, 1969). We will prove optimality with respect to the IMSE of the 4th order kernel in its class, and our conjecture is, that this holds for general orders. Pending a proof of this conjecture, we call this hierarchy of kernels the optimal kernels.

<u>Theorem 6</u>

The 4th order kernel w_4 derived from (26) is optimal with respect to IMSE among all 4th order kernels which exhibit a minimal number of changes of sign.

<u>Proof:</u> Let w_4^* be a symmetric kernel defined by $w_4^*(x) = w_4(x) + \Delta w_4(x)$, satisfying all the requirements of a 4th order kernel, and having the same 4th moment β_4 as w_4 (but not necessarily the same support). The only change of sign to the right of the origin (at some point ρ, say) goes from positive to negative. Otherwise we would have:

$$0 = \int_0^\infty w_4(x)x^2dx = \xi_1^2 \int_0^\rho w_4(x)dx + \xi_2^2 \int_\rho^\infty w_4(x)dx > 0$$

as $\int_0^\infty w_4(x)dx = \frac{1}{2}$, and $\xi_1^2 < \xi_2^2$

In order to establish:

$$\int_{-\infty}^{+\infty} w_4^*(x)^2dx > \int_{-\infty}^{+\infty} w_4(x)^2dx$$

it is sufficient to prove:

$$\int_1^\infty \Delta w_4(x)\, p_4(x)dx \leq 0$$

This follows from the moment conditions, with $w_4(x) = p_4(x)[-1,+1]$:

$$\int_{-\infty}^{+\infty} w_4^*(x)^2dx = \int_{-\infty}^{+\infty} w_4(x)^2dx + \int_{-\infty}^{+\infty} \Delta w_4(x)^2dx$$

$$+ 2\int_-^+ (w_4(x) - p_4(x)) \cdot \Delta w_4(x)dx$$

a) $\rho \leq 1$: The proof is immediate

b) $\rho > 1$: By the mean value theorem of integration we obtain:

$$\int_1^\infty \Delta\, w_4(x)\; p_4(x)dx = \frac{p_4(\xi_1)}{\xi_1^2}\int_1^\rho \Delta\, w_4(x)x^2dx + \frac{p_4(\xi_2)}{\xi_2^2}\int_\rho^\infty \Delta\, w_4(x)x^2dx$$

$$\xi_1 \in (1,\rho)\ ,\ \xi_2 \in (\rho,\infty)$$

We have $\frac{p_4(\xi_2)}{\xi_2^2} \geq \frac{p_4(\xi_1)}{\xi_1^2} \geq 0$ from the following lemma (the proof is left to the reader):

Lemma 3 Let $p(x) = \gamma(x^2 - \alpha_1^2) \ldots (x^2 - \alpha_n^2)\ (\gamma > 0)$ be a symmetric polynomial. Then we have, for $0 \leq j \leq 2n$:

$$\frac{p(x)}{x^j} \text{ is strictly monotone increasing for } |x| > \max_{1\leq i\leq n} |\alpha_i|$$

Since we have:

$$\int_1^\rho \Delta w_4(x)x^2dx \geq 0\ ,\ \int_\rho^\infty \Delta\, w_4(x)x^2dx \leq 0$$

It is sufficient to prove:

$$\int_1^\infty \Delta\, w_4(x)x^2dx \leq 0$$

This latter statement follows from:

$$0 = \int_0^\infty w_4^*(x)x^2dx = \int_0^1 (w_4(x) + \Delta w_4(x))\ x^2dx + \int_1^\infty \Delta\, w_4(x)x^2dx$$

Smoothing methods are primarily variance reducing methods, and bias is a problem we have to cope with. It is thus of interest to obtain minimum variance kernels:

$$(31) \qquad \int v(x)^2 dx = \text{Min!}$$

$$(32) \qquad \int v(x) x^j dx = 1 \qquad j = 0$$

$$= 0 \qquad j = 1, \ldots, k-1$$

<u>Theorem 7</u>

Minimum variance kernels of order k on [-1,+1] are symmetric polynomials of degree (k-2) with (k-2) real roots in (-1,+1). They are unique, and the coefficients are defined by the following system of equations:

$$(33) \qquad 2 \begin{bmatrix} 1 & \frac{1}{3} & \cdots & \frac{1}{k-1} \\ \frac{1}{3} & \frac{1}{5} & \cdots & \frac{1}{k+1} \\ \vdots & \vdots & & \vdots \\ \frac{1}{k-1} & \frac{1}{k+1} & \cdots & \frac{1}{2k-3} \end{bmatrix} \begin{bmatrix} \alpha_0 \\ \alpha_2 \\ \vdots \\ \alpha_{k-2} \end{bmatrix} = \begin{bmatrix} 1 \\ 0 \\ \vdots \\ 0 \end{bmatrix}$$

<u>Proof:</u> Variational calculus applied to the minimum problem (31) renders a polynomial of degree (k-1) at most as a solution; to be integrable such a kernel has to have compact support:

$$V(x) = \sum_{i=0}^{k-1} \alpha_i x^i \qquad \text{on the support } S$$

$$= 0 \qquad \text{outside the support } S$$

We can prove for such kernels of any order k , that they are not only extremal but minimal among all kernels of order k with compact support; by a suitable choice of the support S , the support of a variation δv will be contained in S (otherwise, the problem is ill-posed):

$$\int_S (v(x) + \delta v(x))^2 dx = \int_S v(x)^2 dx + \int_S \delta v(x)^2 dx + 2 \int_S v(x) \; \delta v(x) dx$$

$$> \int_S v(x)^2 dx$$

This follows from the moment conditions applied to δv . We have a unique minimum, and since $v^*(x) = v(-x)$ is also a solution to the minimum problem the kernel has to be symmetric. In particular, we can normalize the symmetric support to $[-1,+1]$. The existence of (k-2) real roots can be deduced from lemma 2. The system of equations (33) follows from the side conditions.

<u>Remark:</u>

Minimum variance kernels automatically fulfill the requirement of a minimal number of sign changes.

The expressions for the asymptotic variance and the asymptotic bias are quite simple:

(34) $$V = \alpha_o$$

(35) $$B = 2 \sum_{i=o}^{k-2} \alpha_i / (i + k + 1)$$

The analytical form for the minimum variance kernels of order k = 2,4,6, as well as the asymptotic variance, the asymptotic bias and MSE/IMSE can be taken from table 2 (in all cases for support [-1,+1]). Figure 2 contains the graphs of these kernels. Asymptotically we do not win much in IMSE by using optimal kernels instead of minimum variance kernels: The ratios of the asymptotic IMSE at the optimal bandwidth are for order k = 2-8 1.060, 1.030, 1.027, 1.021.

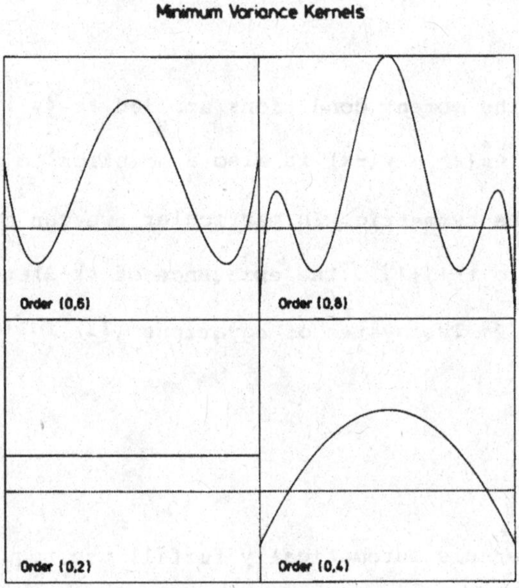

fig. 2: Minimum variance kernels of order 2-8, common min-max-scale, dotted line at zero

The hierarchies of kernels proposed here are of interest beyond the estimation of regression functions: they are optimal with respect to IMSE and variance respectively for the kernel estimation of probability densities, as well as for the nonparametric estimation of spectrum densities via smoothed periodograms.

6. The Boundary Problem

Those cases are rare, where no boundary effects arise due to the validity of a periodic model. Most people familiar with graphic data analysis know the boundary phenomena, as for example the bending of a straight line to a constant towards the end. The arguments below show - somewhat surprisingly - that end effects may dominate the global asymptotic behaviour, when some trivial extension of the estimate (3) to the boundary is used. Note that this problem does not arise for kernel estimation of densities.

To study the asymptotics (as $n \to \infty$, $b(n) \to 0$, $nb(n) \to \infty$) of smoothing estimates at the boundary, we have to consider not a fixed point $t \in [0,b(n))$, $t \in (1-b(n),1]$, but a sequence of points $t(n)$ satisfying:

$$t(n) = q\, b(n)$$
$$q \in [0,1)$$
$$t(n) = 1 - q\, b(n)$$

The derivation of the integral approximations (appendices 1,2) needs minor modifications for a sequence of points. For $t(n) = 1 - q\, b(n)$, the support of the kernel of the estimate (3) that is mapped into the interval $[0,1]$ is $[-q,1]$ instead of $[-1,+1]$ for an interior point. When using the definition (3) as it stands we face a normalization bias.

The "cut-and-normalize" modification omits that part of the kernel lying between -1 and $-q$, and normalizes the kernel between $-q$ and $+1$. The variance is then $O\left(\frac{1}{n\,b(n)}\right)$ and the squared bias $O(b(n)^2)$ (symmetry is lost, and the first moment is different from zero). When integrating over the boundary we obtain $O(b(n)^3)$, whereas the integrated squared bias for the interior is $O(b(n)^{2k})$. The boundary term, therefore, dominates the global asymptotic behaviour even for $k = 2$, and the effect becomes worse for higher order kernels. We ask for the following properties for boundary kernels w_q , $0 \le q < 1$ (compare theorem 4):

(i) The moment conditions (12),(13) have to be fulfilled, and the k-th moment has to be uniformly bounded for $q \in [0,1)$.

(ii) The asymptotic variance has to be uniformly bounded:

$$\int_{-q}^{1} w_q(x)^2 dx < c \quad \forall q$$

(iii) The kernels depend continuously on q , and we have (w = kernel of order k for the interior):

$$w_q \to w \text{ , as } q \to 1$$

We assume w_q to be Lipschitz-continuous of order 1 as a function of x (with the exception of a finite number of points).

Due to requirement (iii) w_q is a well-behaved function of q on the compact interval $[0,1]$. The global asymptotic behaviour is unaffected by the choice of the kernel at the boundary, as long as it fulfills (i)-(iii) (theorem 4); the two following proposals obey the further restriction that the kernel w_q does not have more changes of sign than w has (a positive kernel in particular stays positive at the boundary):

Onesided modification by multiplication with a function g such that the moment conditions (12), (13) and $w_q \to w (q \to 1)$ are fulfilled:

$$w_q(x) = w(x) \qquad 0 \le x \le 1$$
$$= g(x)w(x) \qquad -q \le x < 0$$

The bandwidth shrinks in such a way that the interval of support remains symmetric:

$$w_q(x) = \frac{1}{q}\, w\left(\frac{x}{q}\right) \qquad |x| \le q$$

By definition the requirements (i) and (iii) are satisfied; the variance, however, becomes unbounded as $q \to 0$. In what follows we will give solutions for the hierarchies of minimum variance kernels and of optimal kernels, and for general weight functions for the important case k = 2. A minimum variance kernel of order k at the boundary point q is a solution to the following minimum problem:

$$(34) \qquad \int_{-q}^{1} v_q(x)^2 dx = \text{min!}$$

$$(35) \qquad \int_{-q}^{1} v_q(x) x^j dx = \begin{cases} 1 & j = 0 \\ 0 & j = 1, \ldots, k-1 \end{cases}$$

<u>Theorem 8</u>

There is exactly one minimum variance kernel $v_q(x) = \sum_{i=0}^{k-1} \alpha_i x^i$. The coefficients $\{\alpha_i\}$ are determined by the following system of linear equations:

$$(36)\quad \begin{bmatrix} 1+q & \frac{1}{2}(1-q^2) & \dots & \frac{1}{k}(1-(-q)^k) \\ \frac{1}{2}(1-q^2) & \frac{1}{3}(1+q^3) & \dots & \frac{1}{k+1}(1-(-q)^{k+1}) \\ \vdots & \vdots & & \vdots \\ \frac{1}{k}(1-(-q)^k) & \frac{1}{k+1}(1-(-q)^{k+1}) & \dots & \frac{1}{2k-1}(1-(-q)^{2k-1}) \end{bmatrix} \begin{bmatrix} \alpha_0 \\ \alpha_1 \\ \vdots \\ \alpha_{k-1} \end{bmatrix} = \begin{bmatrix} 1 \\ 0 \\ \vdots \\ 0 \end{bmatrix}$$

The conditions (i)-(iii) are satisfied by the minimum variance kernels, extended to the boundary.

<u>Proof:</u> A variational argument applied to (34), (35) leads to a polynomial of degree up to (k-1) with coefficients determined by (36). To prove that this solution actually attains the minimum, and that any variation strictly increases the variance, the same argument as in theorem 7 can be used. The regularity of the matrix (36) can be proved in the same way as in theorem 5.

Condition (i) is satisfied, the k-th moment being equal to

$$(1 + q^{k+1}) \sum_{i=0}^{k-1} \alpha_i (1-(-q)^{i+k+1})/(i+k+1)$$

The asymptotic variance is equal to α_0 and therefore bounded. By relation (36), the kernels depend continuously on q, and a comparison with § 5 shows that $w_q \to w$ as $q \to 1$.

Boundary	Kernel	Functional form $w_q(x)$	$\int w(x)x^2dx$	$\int w(x)^2dx$	Ratio to cut-and-normalize
q = 1	Min.Var.	$.5$	.3333	.5000	1.000
	Optimal	$.75 - .75\,x^2$	.2000	.6000	1.000
q = 0.8	Min.Var.	$.5761 - .2058\,x$	.2600	(.5556) .5761	1.037
	Optimal 1	$.7742 - .1052\,x - .7742\,x^2 + .1052\,x^3$	.1820	(.6296) .6343	1.007
	Optimal 2	$.7247 - .09145x - .5716\,x^2$	.2000	.6104	0.970
q = 0.6	Min.Var.	$.7422 - .5859\,x$	.1733	(.6250) .7422	1.188
	Optimal 1	$.8776 - .4735\,x - .8776\,x^2 + .4735\,x^3$	.1388	(.7041) .7649	1.087
	Optimal 2	$.6628 - .7690\,x + .4578\,x^2$	.2000	.7544	1.071
q = 0.4	Min.Var.	$1.108 - 1.312x$	.0733	(.7143) 1.108	1.551
	Optimal 1	$1.188 - 1.374x - 1.188\,x^2 + 1.374\,x^3$	.07476	(.8170) 1.115	1.365
	Optimal 2	$.7970 - 3.856x + 4.239\,x^2$	.2000	1.645	2.013
q = 0.2	Min.Var.	$1.944 - 2.778x$	-.0400	(.8333) 1.944	2.333
	Optimal 1	$2.112 - 3.580x - 2.112\,x^2 + 3.580x^3$	-.0098	(.9753) 2.027	2.078
	Optimal 2	$2.639 - 16.67x + 17.36\,x^2$	.2000	6.111	6.266
q = 0.0	Min.Var.	$4.00 - 6.00\,x$	-.1667	(1.000) 4.000	4.000
	Optimal 1	$5.053 - 9.474x - 5.053\,x^2 + 9.474\,x^3$	-.1158	(1.200) 4.498	3.748
	Optimal 2	$15.00 - 72.00x + 66.00\,x^2$	.2000	28.20	23.50

Table 3: Behaviour of kernels of order k=2 at boundary (variance in parantheses: cut-and-normalize method)

The uniform kernel, the minimum variance kernel of order k = 2 for the interior becomes a straight line with slope not equal to zero at the boundary (the slope increasing approximately exponentially towards q=0). For k=4, the minimum variance kernels are displayed for q = 1.0, 0.8, 0.6, 0.4, 0.2, 0.0 in fig. 3.

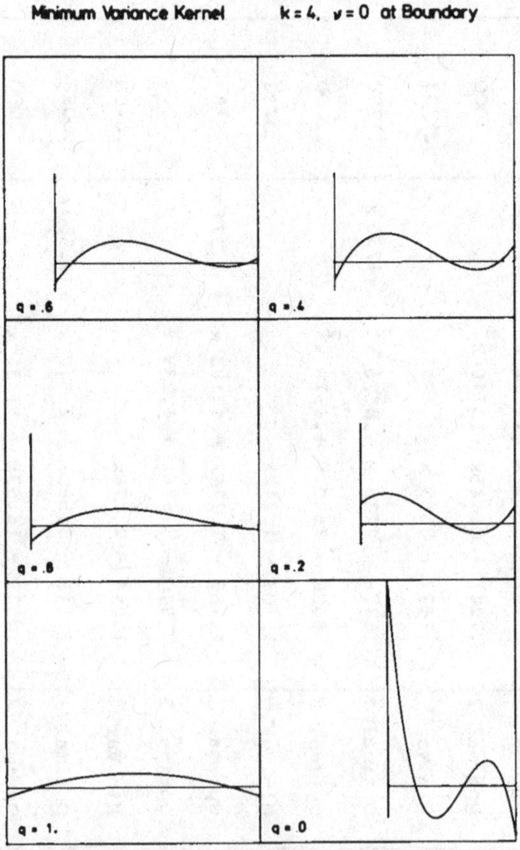

fig. 3: Minimum variance kernels k=4 at the boundary, common min-max-scale dotted line at 0

It is graphically evident (compare also table 3) that the variation within a kernel (and thus the variance) becomes big close to $q = 0$ only. Further tables of the boundary kernels will be given in a contribution devoted to aspects of computation and simulation.

For the hierarchy of optimal kernels, we can propose a multitude of asymptotically valid solutions (in the sense of conditions (i)-(iii)), but no natural continuation of the side conditions except for $k = 2$. For the class of symmetric, positive kernels ($k = 2$) in general we propose to multiply the kernel by a linear function satisfying (i) and (iii) (defined as "optimal 1" in table 3 for the continuation of $w(x) = .75-.75\,x^2$ to w_q). For the continuation of optimal kernels of any order k, we have studied two strategies (both allowing more changes of sign than w):

a) Define the boundary kernel as a polynomial of degree k, $w_q(x) = \sum_{i=o}^{k} \lambda_i(q)x^i$ and determine the coefficients $\lambda_i(q)$ by setting:

$$\int_{-q}^{1} w_q(x)x^j dx = 1 \qquad j = 0$$

$$= 0 \qquad j = 1, \ldots, k-1$$

$$= \beta_q \qquad j = k$$

where β_q depends continuously on q, and $\beta_q \to \beta$ as $q \to 1$ (β = k-th moment of w). The existence and uniqueness of the solution follows by arguments used in § 5.

β) Define the boundary kernel as a polynomial of degree (k+1), $w_q(x) = \sum_{i=0}^{k+1} \gamma_i(q)x^i$ and determine the $\gamma_i(q)$ from the (k+1) equations of α) and by setting $w_q(1) = 0$.

It turns out that the choice of β_q is crucial for the size of the asymptotic variance: Table 3 contains an example of solution α) for k = 2 , by defining $\beta_q = \beta$ (given as "optimal 2"). The increase in variance as q goes from 1 to 0 is very high as compared with the solution obtained by multiplying the kernel w(q = 1) with a linear function; this is even worse for proposal β) with $\beta_q = \beta$. Further results indicate that it is not easy to propose a functional dependence for β_q with good variance properties by trial and error. The cut-and-normalize method is included despite the fact that it does not satisfy condition (i), in order to compare the price we have to pay for condition (i) in term of the variance.

7. Concluding Remarks

It is quite simple to have a coded version for kernel estimates of any order k implemented, much simpler than for smoothing splines. Simulation experiments are urgently needed to check the validity of the asymptotics and to have a comparison with splines. Preliminary Monte Carlo results indicate that the finite sample behaviour of a 4th order kernel is as good or superior to that of the comparable 3rd order spline. To determine the smoothing parameter automatically from the data is of great importance when applying smoothing methods: the definition (3) and the asymptotic theory given in § 3 and § 4 were useful in deriving encouraging results on a cross-validation method to estimate the degree of smoothing from the data.

Acknowledgements: The first author had many fruitful discussions with Werner Stützle and Peter J. Huber on this topic (for which the Zürich longitudinal growth study was a rewarding interdisciplinary background). He thanks both.

R E F E R E N C E S

BENEDETTI, J.K. (1977). On the nonparametric estimation of regression functions. J.Roy.Statist.Soc. B 39, 248-253.

BOCK, R.D., WAINER, H., PETERSEN, A., THISSEN, D., MURRAY, J., ROCHE, A. (1973). A parameterization for individual human growth curves. Human Biology 45, 63-80.

CRAVEN, P., WAHBA, G. (1977). Smoothing noisy data with spline functions Estimating the correct degree of smoothing by the method of generalized cross-validation. Tech.Rep. No. 445, Dept.of Statistics, University of Wisconsin-Madison.

EPANECHNIKOV, V.A. (1969). Nonparametric estimation of a multivariate probability density. Theor.Prob.Appl. 14, 153-158.

GASSER, TH. (1977). General characteristics of the EEG as a signal. EEG Informatic, 37-55, ed. A. RÉMOND, Elsevier/North-Holland Biomedical Press.

LARGO, R.H., STÜTZLE, W., GASSER, TH., HUBER, P.J., PRADER, A. (1978). A description of the adolescent growth spurt using smoothing spline functions. Annals of Human Biology, in print.

PRIESTLEY, M.B., CHAO, M.T. (1972). Nonparametric function fitting. J.Roy.Statist.Soc. B 34, 385-392.

ROSENBLATT, M. (1971). Curve estimates. Ann.Math.Statist. 42, 1815-1841

Appendix 1

The derivation of the integral approximation for the expectation is given for definition type 3 . In this case the proposition is as follows:

$$E(\hat{\mu}_{n,3}(t)) = \int_{-\tau}^{\tau} w(x)\ \mu(t-xb(n)dx + O\left(\frac{1}{n^{\gamma_\mu}}\right)$$

The argument is based on the mean value theorem of integration, using the compactness of support of the kernel w and its Lipschitz continuity of order γ_μ :

$$\left|E(\hat{\mu}_{n,3}(t)) - \frac{1}{b(n)} \int_0^1 w\left(\frac{t-x}{b(n)}\right) \mu(x)dx\right|$$

$$\leq \frac{1}{b(n)} \sum_{j=1}^{n} \left| \int_{s_{j-1}^{(n)}}^{s_j^{(n)}} w\left(\frac{t-x}{b(n)}\right) (\mu(t_j^{(n)} - \mu(x))dx\right|$$

$$\leq \frac{\max_x |w(x)|}{b(n)} \sum_{j\in\mathcal{J}} | s_j^{(n)} - s_{j-1}^{(n)} | \ | \mu(t_j^{(n)} - \mu(\xi_j^{(n)}) |$$

where $\xi_i^{(n)}$ are suitable mean values and $|\mathcal{J}| = O(nb(n))$. Using (8), we obtain for some constant α :

$$\leq \frac{\max_x |w(x)|}{b(n)} \mid \mid \frac{\alpha}{n} \left(\frac{\alpha}{n}\right)^{\gamma_\mu} \cdot |\mathcal{J}|$$

$$= O\left(\frac{1}{n^{\gamma_\mu}}\right)$$

Appendix 2

For the integral approximation of the variance of estimate 3 , we have the following proposition:

$$\operatorname{Var}(\hat{\mu}_{n,3}(t)) = \frac{\sigma^2}{nb(n)} \int_{-\tau}^{\tau} w(x)^2 dx + O\left(\frac{1}{n^{1+\gamma_w} b(n)^{1+\gamma_w}} + \frac{1}{n^{\delta} b(n)}\right)$$

$$\left|\operatorname{Var}(\hat{\mu}_{n,3}(t)) - \frac{\sigma^2}{nb(n)^2} \int_0^1 w\left(\frac{t-x}{b(n)}\right)^2 dx\right|$$

$$\leq \frac{\sigma^2}{b(n)^2} \left| \sum_{j=1}^{n} \left(\int_{s_{j-1}^{(n)}}^{s_j^{(n)}} w\left(\frac{t-x}{b(n)}\right) dx \right)^2 - \frac{1}{n} \int_0^1 w\left(\frac{t-x}{b(n)}\right)^2 dx \right|$$

$$\leq \frac{\sigma^2}{b(n)^2} \sum_{j=1}^{n} \left| (s_j^{(n)} - s_{j-1}^{(n)})^2 \, w\left(\frac{t-\xi_j^{(n)}}{b(n)}\right)^2 - \frac{1}{n} \left(s_j^{(n)} - s_{j-1}^{(n)}\right) w\left(\frac{t-\theta_j^{(n)}}{b(n)}\right)\right.$$

(where $\xi_j^{(n)}$, $\theta_j^{(n)}$, $j = 1, \ldots, n$, are such that $s_{j-1}^{(n)} \leq \xi_j^{(n)}, \theta_j^{(n)} \leq s_j^{(n)}$

Using (8), and $|\mathfrak{J}| = O(nb(n))$, we obtain for some constant α :

$$\leq \frac{\sigma^2}{b(n)^2} \frac{\alpha}{n} \sum_{j=1}^{n} \left| (s_j^{(n)} - s_{j-1}^{(n)} - \frac{1}{n}) \, w\left(\frac{t-\xi_j^{(n)}}{b(n)}\right)^2 \right.$$

$$\left. -\frac{1}{n} \left(w\left(\frac{t-\theta_j^{(n)}}{b(n)}\right)^2 - w\left(\frac{t-\xi_j^{(n)}}{b(n)}\right)^2 \right. \right|$$

$$\leq \frac{\sigma^2}{b(n)^2} \frac{\alpha}{n} \left(\max_x w(x)^2 \sum_{j\in\mathfrak{J}} \left| s_j^{(n)} - s_{j-1}^{(n)} - \frac{1}{n} \right| \right.$$

$$\left. + \frac{1}{n} \sum_{j\in\mathfrak{J}} \left| c_{\gamma_w} \left(\frac{\theta_j^{(n)} - \xi_j^{(n)}}{b(n)}\right)^{\gamma_w} \right| \right)$$

$$\leq O\left(\frac{1}{n^{\delta} b(n)} + \frac{1}{n^{1+\gamma_w} b(n)^{1+\gamma_w}}\right)$$

TOTAL LEAST SQUARES

Gene H. Golub* and Charles Van Loan.

In many data-fitting situations, we are given an $m \times n$ matrix A, a vector $\underset{\sim}{b}$ with m components and we wish to determine a vector $\hat{\underset{\sim}{x}}$ so that

$$\|D(\underset{\sim}{b} - A\underset{\sim}{x})\|_2 = \min! \tag{1}$$

where D is a diagonal matrix and $\|.\|_2$ indicate the euclidean length of the vector. It is well known that $\hat{\underset{\sim}{x}}$, the solution to (1), satisfies the system of equations

$$A^T D^2 A \hat{\underset{\sim}{x}} = A^T D^2 \underset{\sim}{b} . \tag{2}$$

In this problem, there is a tacit assumption that the errors are confined to the "observations" $\underset{\sim}{b}$. Now let us recast the problem as follows:

determine that $\hat{\underset{\sim}{x}}$ which satisfies

$$A \hat{\underset{\sim}{x}} = \underset{\sim}{b} + \underset{\sim}{r} \tag{3}$$

and for which

$$\|D \underset{\sim}{r}\|_2 = \min! \tag{4}$$

In this paper we assume that there is error in the "data" A. Then it is convenient to consider the problem of determining $\hat{\underset{\sim}{x}}$ which satisfies

$$(A + E) \hat{\underset{\sim}{x}} = \underset{\sim}{b} + \underset{\sim}{r} \tag{5}$$

and for which

$$\|D(E \vdots \underset{\sim}{r}) T\|_F \tag{6}$$

where $D = \mathrm{diag}(d_1,\dots,d_m)$, $T = \mathrm{diag}(t_1,\dots,t_{n+1})$, $(d_i > 0,\ t_i > 0$ for all $i)$ and $\|\cdot\|_F$ indicates the Frobenius norm of the matrix; viz $\|B\|_F^2 = \sum_{i,j} |b_{ij}|^2$. In many situations, $T = \mathrm{diag}(1,1,\dots,1,\lambda)$; this is closely analogous to ridge regression. Once a minimizing $(E \vdots \underset{\sim}{r})$ is found subject to the constraints, then there exists a vector $\hat{\underset{\sim}{x}}$ such that

$$(A + E)\hat{\underset{\sim}{x}} = \underset{\sim}{b} + \underset{\sim}{r} .$$

---

* This author has been in part supported to DOE contract No. DE-AS03-76SF00326 and United States Army Research grant No. DAAG 29-78-G-0179.

We refer to this problem as the "total least squares" (TLS) problem. The TLS problem may not have a solution. For example, if

$$A = \begin{bmatrix} 1 & 0 \\ 0 & 0 \end{bmatrix}, \quad \underset{\sim}{b} = \begin{bmatrix} 1 \\ 1 \end{bmatrix}, \quad S = T = I_2$$

and

$$E_\epsilon = \begin{bmatrix} 0 & 0 \\ 0 & \epsilon \end{bmatrix}$$

then for $\epsilon > 0$ $(A + E_\epsilon)\, \underset{\sim}{x}_\epsilon = \underset{\sim}{b}$. Clearly there is no smallest value of $\|(E\vdots\underset{\sim}{r})\|_F$ for which (5) is satisfied. We shall give general conditions under which it is impossible to determine a solution.

A generalization of the TLS problem results if we allow multiple right hand sides. Hence we wish to find $\hat{X}$ so that

$$(A + E)\, \hat{X} = B + R \tag{7}$$

where B is an $m \times k$ matrix and for which

$$\|D(E\vdots R)T\|_F = \min! \tag{8}$$

where $T = \mathrm{diag}(t_1, t_2, \ldots, t_{n+k})$.

We shall first consider the case where $k = 1$. We re-write (5) as follows:

$$(A\vdots\underset{\sim}{b}) \begin{pmatrix} \hat{\underset{\sim}{x}} \\ -1 \end{pmatrix} + (E\vdots\underset{\sim}{r}) \begin{pmatrix} \hat{\underset{\sim}{x}} \\ -1 \end{pmatrix} = \underset{\sim}{0}$$

so that

$$\{D(A\vdots\underset{\sim}{b})T + D(E\vdots\underset{\sim}{r})T\}\, T^{-1} \begin{pmatrix} \hat{\underset{\sim}{x}} \\ -1 \end{pmatrix} = \underset{\sim}{0} . \tag{9}$$

Writing

$$D(A\vdots\underset{\sim}{b})T = C$$

$$D(E\vdots\underset{\sim}{r})T = \Delta$$

and

$$T^{-1} \begin{pmatrix} \hat{\underset{\sim}{x}} \\ -1 \end{pmatrix} = \underset{\sim}{y} ,$$

(9) is equivalent to

$$(C + \Delta)\, \underset{\sim}{y} = \underset{\sim}{0} . \tag{10}$$

In order to find a vector $\underset{\sim}{y}$ which satisfies the homogeneous equation (10), we need to find a matrix Δ so that

$$\mathrm{rank}(C + \Delta) < \min\,(m, n+1) . \tag{11}$$

Furthermore, in accordance with (6),

$$\|\Delta\|_F = \min! \tag{12}$$

The solution to (11) and (12) is given in terms of the singular value decomposition (SVD) of C . (In what follows, we assume $m \geq n+1$.) It is well known (cf. [5]) that every $m \times (n+1)$ matrix can be written as follows:

$$C = U \Sigma V^T \tag{13}$$

where

$$U^T U = I_m ,$$

$$V^T V = I_{n+1}$$

and

$$\Sigma = \begin{pmatrix} \sigma_1 & & \bigcirc \\ & \ddots & \\ \bigcirc & & \sigma_{n+1} \\ \cdots & \cdots & \cdots \\ & \bigcirc & \end{pmatrix}_{m \times (n+1)} .$$

The matrix Σ consists of the singular values of C with $\sigma_1 \geq \sigma_2 \geq \cdots \geq \sigma_{n+1} \geq 0$ and $U(V)$ is the matrix of eigenvectors of $CC^T (C^T C)$. The singular values are the non-negative square roots of the eigenvalues of $C^T C$. Hence the number of non-zero singular values of C equals the rank of C . Let us write $U = (\underset{\sim}{u}_1, \ldots, \underset{\sim}{u}_m)$, $V = (\underset{\sim}{v}_1, \ldots, \underset{\sim}{v}_{n+1})$; hence if $\sigma_i = \sigma_{i+1}$, $\underset{\sim}{u}_i$, $\underset{\sim}{u}_{i+1}$, $\underset{\sim}{v}_i$, $\underset{\sim}{v}_{i+1}$ are not unique. Now consider the following problem.

Given C , find the matrix C_p of rank p , so that

$$\|C - C_p\|_F = \min!$$

The solution can be given in terms of the SVD of C (see [2] for details). Namely

$$C_p = U \Sigma_p V^T \tag{14}$$

where

$$\Sigma_p = \begin{pmatrix} \sigma_1 & & & & \bigcirc \\ & \ddots & & & \\ & & \sigma_p & & \\ & \bigcirc & & 0 & \\ & & & & \ddots \\ & & & & & 0 \\ \cdots & \cdots & \cdots & \cdots \\ & & \bigcirc & & \end{pmatrix}$$

so that

$$C - C_p = \sum_{i=p+1}^{n+1} \sigma_i \underset{\sim}{u}_i \underset{\sim}{v}_i^T . \qquad (15)$$

The solution to our problem then is given by the matrix

$$C_n = U \Sigma_n V^T$$

and hence

$$\Delta = -\sigma_{n+1} \underset{\sim}{u}_{n+1} \underset{\sim}{v}_{n+1}^T . \qquad (16)$$

The solution is unique when $\sigma_n > \sigma_{n+1}$. Later, we shall consider the situation when the smallest singular value is not unique. Now let

$$\underset{\sim}{u}_{n+1} \equiv \underset{\sim}{u} , \quad \underset{\sim}{v}_{n+1} \equiv \underset{\sim}{v} ;$$

so that

$$C_n \underset{\sim}{v} = \underset{\sim}{0} .$$

Recall that

$$C = D(A \vdots \underset{\sim}{b})T .$$

Thus

$$T^{-1} \begin{pmatrix} \hat{\underset{\sim}{x}} \\ -1 \end{pmatrix} = \alpha \chi .$$

Providing

$$v_{n+1} \neq 0 ,$$

$$\hat{x}_i = \alpha t_i v_i \qquad (17)$$

with

$$\alpha = -(v_{n+1} t_{n+1})^{-1} . \qquad (18)$$

As we have noted, the solution is not unique when

$$\sigma_{\ell+1} = \dots = \sigma_{n+1} > 0 .$$

Let us partition the matrix V as follows:

$$V = (\underset{\sim}{v}_1, \underset{\sim}{v}_2, \dots, \underset{\sim}{v}_\ell \vdots \underset{\sim}{v}_{\ell+1}, \dots, \underset{\sim}{v}_{n+1})$$

$$\equiv (V_1 \vdots V_2) .$$

Thus for any $\{c_i\}_{i=1}^{n+1-\ell}$ such that $\sum_i |c_i| > 0$,

$$T^{-1}\begin{pmatrix}\hat{\underset{\sim}{x}}\\ -1\end{pmatrix} = \alpha(c_1\underset{\sim}{v}_{\ell+1} + c_2\underset{\sim}{v}_{\ell+2} + \ldots + c_{n+1-\ell}\underset{\sim}{v}_{n+1})$$

and hence

$$x_i = \alpha t_i v_i \qquad (i = 1,2,\ldots,n)$$

with

$$\alpha = -(t_{n+1}(c_1 v_{n+1,\ell+1} + \ldots + c_{n+1-\ell}v_{n+1,n+1}))^{-1} .$$

We wish to construct the unique vector $\hat{\underset{\sim}{x}}$ so that

$$\|T^{-1}\hat{\underset{\sim}{x}}\|_2^2 = \sum_{i=1}^{n} \frac{\hat{x}_i^2}{t_i^2} = \min!$$

It is simple to construct this solution. Let

$$\underset{\sim}{w}^T = (v_{n+1,\ell+1}, v_{n+1,\ell+2}, \ldots, v_{n+1,n+1}) ;$$

we construct a matrix Q so that

$$Q\underset{\sim}{w} = \|\underset{\sim}{w}\|_2\, \underset{\sim}{e}_1$$

where

$$\underset{\sim}{e}_1^T = (1,0,\ldots,0) \quad \text{and} \quad Q^TQ = I$$

The matrix Q is easy to construct as a Householder matrix (cf. [1]), viz $Q = I - 2\,\underset{\sim}{z}\,\underset{\sim}{z}^T$ with $\|\underset{\sim}{z}\|_2 = 1$. Then

$$V_2Q \equiv \begin{pmatrix} \underset{\sim}{v} & /////// \\ & \underset{\sim}{0}^T \end{pmatrix} ,$$

and the components of $\underset{\sim}{v}$ can be used to construct the solution as described by (17) and (18).

As we noted, earlier, the solution to the TLS problem does not always exist. In particular if $\text{rank}(A) < n$ and $b \notin \text{Ran}(A)$,

$$C\underset{\sim}{v} = \underset{\sim}{0}$$

with

$$v_{n+1} = 0 .$$

Hence no solution exists in this case. If, however, $\underset{\sim}{b} \in \text{Ran}(A)$, $C\underset{\sim}{v} = \underset{\sim}{0}$ with

$$v_{n+1,j} \neq 0 \quad \text{for some} \quad j = \ell+1,\ldots,n+1, \quad \text{a solution exists.}$$

In this situation the solution may not be unique.

Now let us assume rank $(A) = n$. Recall that

$$C^TC = V\,\Sigma^T\Sigma\,V^T$$

so that the smallest eigenvalue of C^TC is the square of the smallest singular value of C . For notational convenience we define the following:

$$D\,A\,T_1 = \tilde{A}\,, \quad D\underset{\sim}{b}\,t_{n+1} = \tilde{\underset{\sim}{b}}$$

where

$$T_1 = \text{diag}(t_1,\ldots,t_n)\,.$$

Hence

$$C^TC = \begin{pmatrix} \tilde{A}^T\tilde{A} & \tilde{A}^T\tilde{\underset{\sim}{b}} \\ \tilde{\underset{\sim}{b}}^T\tilde{A} & \tilde{\underset{\sim}{b}}^T\tilde{\underset{\sim}{b}} \end{pmatrix}.$$

Thus if $\tilde{A}^T\tilde{\underset{\sim}{b}} = \underset{\sim}{0}$ and $\lambda_{min}(\tilde{A}^T\tilde{A}) < \tilde{\underset{\sim}{b}}^T\tilde{\underset{\sim}{b}}$ ($\lambda_{min}(\tilde{A}^T\tilde{A})$ indicates the smallest eigenvalue of $\tilde{A}^T\tilde{A}$),

$$v_{n+1} = 0\;,$$

and hence no solution exists. (Note that in the usual least squares situation $\tilde{A}^T\tilde{\underset{\sim}{b}} = \underset{\sim}{0}$ implies $\hat{\underset{\sim}{x}} = \underset{\sim}{0}$.)

There is a close connection between the TLS problem and ridge regression. Recall that in ridge regression, one solves the system of equations

$$(\tilde{A}^T\tilde{A} + \mu I)\hat{\underset{\sim}{x}} = \tilde{A}^T\tilde{\underset{\sim}{b}}$$

with $\mu \geq 0$.

Now consider the TLS problem where

$$T_1 = I_n \text{ and } t_{n+1} = \lambda\,.$$

The TLS solution is found by computing the smallest eigenvalue and corresponding eigenvector of

$$C^TC = \begin{pmatrix} \tilde{A}^T\tilde{A} & \lambda\tilde{A}^T\tilde{\underset{\sim}{b}} \\ \lambda\tilde{\underset{\sim}{b}}^T\tilde{A} & \lambda^2\tilde{\underset{\sim}{b}}^T\tilde{\underset{\sim}{b}} \end{pmatrix}$$

A short manipulation shows that

$$(\tilde{A}^T\tilde{A} - \sigma^2_{n+1}(\lambda)\;I)\hat{\underset{\sim}{x}} = \tilde{A}^T\tilde{\underset{\sim}{b}}\;. \tag{19}$$

where $\sigma_{n+1}(\lambda)$ is the smallest singular value of C . Thus for the TLS problem we shift the spectrum of $\tilde{A}^T\tilde{A}$ downward and this is in sharp distinction to the ridge regression situation. Note that

$$\sigma^2_{n+1}(\lambda) < \sigma^2_{min}(\tilde{A}^T\tilde{A})$$

when $\tilde{A}^T\underset{\sim}{b} \neq \underset{\sim}{0}$. So that the system (19) is always positive definite when $\tilde{A}^T\tilde{A}$ is. We observe that $\sigma^2_{n+1}(\lambda) \to 0$ as $\lambda \to 0$.

Finally, we wish to study a method for computing σ_{min} as a function of λ . Let us write

$$\tilde{A}^T\tilde{A} = P M P^T$$

where P is the set of eigenvectors of $\tilde{A}^T\tilde{A}$ and $M = \text{diag}(\mu_1^2,\ldots,\mu_n^2)$, the eigenvalues of $\tilde{A}^T\tilde{A}$.

Let

$$\underset{\sim}{g} = P^T\tilde{A}^T\underset{\sim}{b} \quad , \quad h^2 = \underset{\sim}{b}^T\underset{\sim}{b} \ .$$

A short manipulation (see [4] for details), shows

$$\frac{\sigma^2}{\lambda^2} + \sum_{i=1}^{m} \frac{g_i^2}{\mu_i^2 - \sigma^2} = h^2 \tag{20}$$

The numerical solution of (20) for the smallest singular value can be accomplished quite cheaply by a variety of zero finding techniques.

Finally, we are concerned with the multivariate situation. We wish to examine the problem when $\underset{\sim}{b}$ is replaced by a set of k vectors. Much of the analysis goes through as described above. In this situation, we replace (10) by

$$(C + \Delta)Y = 0$$

where

$$Y = T^{-1}\begin{pmatrix} \hat{X} \\ -I_k \end{pmatrix}$$

and T is a diagonal matrix of dimension $n + k$. The solution Y can be obtained from the SVD of C :

$$C = U \Sigma V^T .$$

We seek the matrix C_{n-k+1} which is the closest matrix of rank $(n-k-1)$ to the original matrix C . Then if

$$\sigma_{n-k+1} > \sigma_{n-k+2} \ , \tag{21}$$

and if

$$V = \begin{pmatrix} V_{11} & V_{12} \\ V_{21} & V_{22} \end{pmatrix}$$

where V_{22} is a $k \times k$ matrix with $\det V_{22} \neq 0$,

$$\hat{X} = - V_{12} V_{22}^{-1} T_2^{-1}$$

where

$$T_2 = \operatorname{diag}(t_{n+1}, \ldots, t_{n+k}) .$$

The situation where (21) does not hold will be discussed in a future paper.

Computer Science Department
Stanford University
Stanford, CA 94305
USA

Computer Science Department
Cornell University
Ithaca, NY 14853
USA

References

[1] Golub, G. H., "Numerical methods for solving linear least squares problems," Numer. Math., 7, 206-16 (1965).

[2] Golub, G. H. and W. Kahan, "Calculating the singular values and pseudo-inverse of a matrix." SIAM J. Numer. Anal. Ser. B., 2, 205-24 (1965).

[3] Golub, G. H. and C. Reinsch, "Singular value decomposition and least squares solutions," Numer. Math. 14, 403-20 (1970).

[4] Golub, G. H., "Some modified matrix eigenvalue problems," SIAM Rev. 15, 318-335 (1973).

[5] Lanczos, C., Linear Differential Operators, Van Nostrand, London, 1961, Chapter 3

SOME THEORETICAL RESULTS ON TUKEY'S 3R SMOOTHER

C.L. Mallows
Bell Laboratories
Murray Hill, NJ 07974/USA

ABSTRACT

Recently Tukey has proposed several non-linear smoothers for time series, which have some properties that make them preferable in some ways to linear filters. We discuss these properties, and give some detailed results for one of these smoothers.

1. Introduction

In recent years, notably in [7], J. W. Tukey has drawn attention to the problem of smoothing when the data may be contaminated by outliers, and may exhibit abrupt changes in level. He has proposed several non-linear algorithms that are simultaneously (i) resistant to occasional outliers and (ii) responsive to abrupt changes; an example is "3R", in which the running-median-of-length-three smoother (which is called "3") is applied repeatedly, until no further change occurs. Thus if for all integral t we have

$$X_t^0 =_{\text{def}} X_t$$

then

$$3(X^c)_t =_{\text{def}} X_t^{c+1} =_{\text{def}} \text{med}\left(X_{t-1}^c, X_t^c, X_{t+1}^c\right)$$

and

$$3R(X)_t =_{\text{def}} \lim_{c\to\infty} X_t^c .$$

Very few properties of such smoothers are known; here we shall list some desiderata for non-linear smoothers generally, and shall present some detailed results for 3R.

2. Some non-linear smoothers

Let S denote a general smoother; the result of applying S to a given time-series $\underline{X} = \{X_t : -<t<\infty\}$ is a new series $S(\underline{X}) = \{S(\underline{X})_t : -\infty<t<\infty\}$ We shall ignore end-effects, though in practice detailed rules need to be given for handling the ends of the given series, and these choices may be critically important (particularly when a smoother is being used for prediction). All the smoothers we shall consider here are

(i) scale- and translation-invariant : $S(a\underline{X}+b)_t = aS(\underline{X})_t+b$

(ii) stationary: : $S(B^k\underline{X})_t = S(\underline{X})_{t-k}$

where B is the back-shift operator, $(B\underline{X})_t =_{\text{def}} X_{t-1}$; and

(iii) centered : $S(\underline{0}) = \underline{0}$.

Also most of them

(iv) pass a linear trend : $S(\alpha+\beta t) = \alpha+\beta t$.

However they are not all invariant under linear detrending (in general $S(\{X_t+\beta t\})_t \neq S(\underline{X})_t+\beta t$); and more generally they are not linear:

$S(\underline{X}+\underline{Y})_t \neq S(\underline{X})_t+S(\underline{Y})_t$.

Many smoothers satisfying (i)-(iv) can be written down; clearly if S_1, S_2 are two such, the compount smoother $S_1(S_2(\cdot))$ is again of this type. A linear filter with coefficients $\{a_n\}$ (output $\Sigma a_u X_{t-u}$) will satisfy (i)-(iv) provided $\Sigma a_u = 1$, $\Sigma u a_u = 0$. Tukey has considered a large number of smoothers built out of elementary operations such as

"m": running median of length m (m odd or even)

"H" ("hanning"): a linear filter, with coefficients (1/4, 1/2, 1/4).

"S" ("splitting"): each flat local maximum or minimum of length 2 ("mesas," and "dales") is regarded as split in two, and some end-of-series rule is applied on each side of the split.

"T" (twicing): If S is a smoother, "S twice" is defined by $ST(\underline{X})_t =_{\text{def}} S(\underline{X})_t + S(X_t - S(\underline{X})_t)$. (If S is linear and passes polynomials of degree $k \geq 0$, ST passes polynomials of degree 2k+1).

Some other families of smoothers are as follows.

M-type: by analogy with Huber's estimates in the location problem, $S(\underline{X})_t$ satisfies $0 = \Sigma a_i \psi(X_{t+i} - S(\underline{X})_t)$ where ψ is an odd function, perhaps monotone.

L-type: $S(\underline{X})_t =_{\text{def}} \int x \, b(F_t(x)dF_t(x)$ where $F_t(x)$ is a local weighted empiric distribution,

$$F_t(x) =_{\text{def}} \sum_i a_i I\{X_{t+i} < x\}$$

and b satisfies $\int_0^1 b(p)dp = 1$.

Robustified AR: $S(\underline{X})_t$ is defined recursively by

$$S(\underline{X})_t =_{\text{def}} \hat{S}_t + s\psi(s^{-1}(X_t - \hat{S}_t))$$

where s is a (robust, possibly local) estimate of the scale of $S(\underline{X})$, $\hat{S}_t$ is a one-step-ahead predictor based on past values of $S(\underline{X})$:

$$\hat{S}_t =_{\text{def}} \sum_{i \geq 1} a_i \hat{S}(\underline{X})_{t-i} ,$$

and $\hat{a}_1, \hat{a}_2, \ldots$, are (robust, possibly local) estimates of autoregressive coefficients (see [5]).

3. What should a good smoother do?

While the properties of individual algorithms can and should be investigated in detail, other immediate needs are for discussion of desiderata for such algorithms, and of development of methods for comparing their properties. Among the properties of a smoother S that might be deemed desirable, depending on the application one has in mind one can list the following:

(a) If X is a stationary Gaussian process, S should be (nearly) linear;

(b) S should be resistant to occasional outliers;

(c) S should be responsive to abrupt changes in level;

(d) When used in curve-fitting, a simple formula for the bias should be available;

(e) More generally, the properties of S should be transparent;

(f) S should be close to optimum (relative to some "central" speci fication, or throughout some neighborhood)

(g) generalizing (c), S should adapt to changes other than simply in level, such as in marginal distribution and covariance structure;

(h) S should be adjustable.

(Items (d) through (h) were suggested by luminaries of the Workshop on Smoothing Techniques for Curve Estimation, held at Heidelberg, April 2-4 1979, to whom I am grateful.)

Lest the above list should appear much too ambitious, let us recall that linear filters already enjoy many of these properties, in fact all except (b) and (c); (g) can be said to be satisfied since when a linear filter is applied to a stationary process the effect is simply to multiply the spectrum of the process by the (squared) transfer function, which depends not at all on the marginal distribution and covariance structure of the process being smoother. However linear filters do not enjoy property (b), and have only limited success in responding to abrupt changes in level.

On the other hand, smoothers that do have properties (b) and (c) are easily constructed; a simple running median is one such. There is thus an immediate challenge, to design smoothers that retain, as far as possible, the pleasant properties of linear filters while achieving also (b) and (c).

A preliminary challenge is to develop ways of measuring the degree to which a given smoother achieves each of its competing objectives. More specifically, one can search for ways of measuring linearity, resistance and responsiveness. It is desirable that these measures should be independent, as far as possible, of the specification of the process being smoothed (as is the case for linear filters). Some general progress in these directions has been made, and is reported in [3] and [4]. In the next section we shall derive some properties of the smoother 3R, and shall then relate these to the general theory.

4. Some properties of 3R

The iterated-running-three-median smoother 3R defined in section 2 above has been vigorously advocated by Tukey in [7]. However he does not give any formal analysis of the properties of this smoother beyond pointing out that it will produce a series that consists of weakly monotone stretches separated by flat maxima (of length at least 2) and flat minima (of length at least 2). (Let $\mathcal{M}$ denote the set of such functions.)

We now present several observations that throw some light on the behavior of this smoother.

Obs. 1. If $x \leq y$ then $x \leq \mathrm{med}(x,y,z) \leq y$.

Obs. 2. If $3(\underline{X})_t = X_t$, then $3R(\underline{X})_t = X_t$.

Proof W.l.o.g. we may assume $X_{t-1} \leq X_t \leq X_{t+1}$. By Obs. 1, $3(\underline{X})_{t-1} \leq X_t = 3(\underline{X})_t \leq 3(X)_{t+1}$ so that, by induction, for all c $3^c(\underline{X})_{t-1} \leq X_t = 3^c(\underline{X})_t \leq 3^c(X)_{t+1}$.

Obs. 2 provides a simple rule for terminating the inductive computation of $3R(\underline{X})$.

Obs. 3. For some $\underline{X}$, $3R(\underline{X})$ is not well-defined; for example if $X_t = (-1)^t$ then $3^k(\underline{X})_t = (-1)^{k+t}$ which does not converge as $k\to\infty$. In theoretical calculations we must be careful to check that such cases have probability zero.

Obs. 4. The value of $3R(\underline{X})_t$ (if it exists) must equal X_{t+k} for some k; k can be indefinitely large with positive probability under reasonable specifications, for example if $\{X_t\}$ are independent and identically distributed.

Obs. 5. The output of 3R is not necessarily the function $\underline{m}$ in M that best fits the data in the L_1 or L_2 sense, even if $\underline{m}$ is further restricted so that for each t, m_t must equal X_{t+k} for some k (as is the case for 3R).

Proof Suppose $\underline{X}$ is

$$\ldots,3,3,3,5,1,8,4,6,2,7,7,7,\ldots .$$

Then $3R(\underline{X})$ is

$$\ldots,3,3,3,3,4,4,5,6,6,7,7,7,\ldots ,$$

whereas the sequence

$$\ldots,3,3,3,3,4,4,4,6,6,7,7,7,\ldots$$

provides a closer fit to $\underline{X}$.

Obs. 6. The result of using 3R to smooth a pure sinusoid with integral period $p\left(X_t = a \sin\left(\frac{2\pi t}{p} + \phi\right)\right)$ is a periodic function, not centered or sinusoidal in general, with an amplitude that, for some values of p, depends crucially on the phase ϕ. (These effects were noticed by Velleman [8]).

Proof (i) If $p = 3$ and $-\pi/6 \leq \phi \leq \pi/6$, $3R(\underline{X})_t = \sin \phi$ for all t.

(ii) If $p = 5$ and $\phi = 0$, $3R(\underline{X})_t = (0,q,q,-q,-q)$ for $t = 0,1,2,3,4$ where $q = \sin(\pi/5)$; these values are not sinusoidal.

(iii) If $p = 4$ and $\phi = 0$, $3R(\underline{X})_t = 0$ for all t; if $p = 4$ and $\phi = \pi/4$, $3R(\underline{X})_t = X_t$ for all t.

Obs. 7. Suppose $\underline{X}$ has a stationary probability specification, and that this specification is "mixing" in one of the senses listed below. Because of Obs. 4, it is a non-trivial question whether or not $3R(\underline{X})$ is "mixing" in the same senses. To state our results we must introduce some notation. Let A, A' be the sigma-fields induced by $\{X_t : t \leq 0\}$ and $\{X_t : t > 1\}$ respectively; for general sets A, A' in A, A' respectively and for $n \geq 0$ set

$$P = \Pr\{A\}, \quad P' = \Pr\{A'\}, \quad P_n = \Pr\{A \cap B^{-n} A'\}$$

(B is the back-shift operator, as before). The following concepts are standard: $\underline{X}$ is

ergodic if $\forall A, A'$, $n^{-1} \sum_{k=1}^{n} (P_k - PP') \to 0$.

weak mixing if $\forall A, A'$, $n^{-1} \sum_{k=1}^{n} |P_k - PP'| \to 0$.

(strong) mixing if $\forall A,A', P_n-PP' \to 0$.

uniform mixing if $\sup_{A,A'} |P_n-PP'| \to 0$.

A minor variation on a definition of Blum, Hanson and Koopmans [1] is: X is *-mixing if $\sup|P_n-PP'|/PP' \to 0$, where the supremum is over all A,A' such that $PP' > 0$.

It is not difficult to prove that if $\underline{X}$ is ergodic (resp. weak mixing, (strong) mixing), then $\underline{S} = 3R(\underline{X})$ is also ergodic (resp. weak mixing, (strong) mixing). However to prove that $\underline{S}$ is uniform mixing it seems to be necessary to assume that $\underline{X}$ is *-mixing. However, I have not been able to construct an example of a uniform-mixing $\underline{X}$ for which $\underline{S}$ is not uniform-mixing; I do not know whether $\underline{S}$ must be *-mixing whenever $\underline{X}$ is. S. P. Lloyd was very helpful in clarifying these relationships.

Obs. 8. The following distributional results are derived in the Appendix. If $\underline{S} = 3R(\underline{X})$ where $\underline{X}$ is a sequence of independent and identically distributed random variables with $\Pr\{X_t \le x\} = F(x)$, then

$$\Pr\{S_t \le x\} = \frac{2F^2-F^3}{1-F+F^2} \qquad (F=F(x)) \tag{4.1}$$

$$= d_1F_{2:3}(x) + d_2F_{3:5}(x) + \ldots + d_kF_{k+1:2k+1} + \ldots \tag{4.2}$$

where $F_{a:b}$ is the distribution of the a-th order-statistic in a sample of size b from F: $F_{a:b} = \int_0^F b\binom{b-1}{a-1}u^{a-1}(1-u)^{b-a}du$, and $d_j = \binom{2j-1}{j-1}^{-1} - \binom{2j+1}{j-1}^{-1}$. (The coefficients d_j are most easily derived by differentiating both (4.1) and (4.2), but the expansion is valid even if F does not have a density.) Thus $d_1 = 2/3$, $d_2 = 7/30$, $d_3 = 1/14$, etc. Further, if C_t is the smallest index $c \ge 1$ such that $3^c(\underline{X})_t = 3^{c-1}(\underline{X})_t$ (so that by Obs. 2 $S_t = 3^c(\underline{X})_t$ and the iteration can be terminated), and if F is continuous at x, then

$$\Pr\{S_t \in dx, S_t = X_{t+k}, C_t = c\}$$
$$= \begin{cases} 2F(1-F)dF(x) & c=1,\ k=0 \\ (1+2F-2F^2)(F(1-F))^{c-1}dF(x) & c\ge 2,\ k=1-c,3-c,\ldots,c-1 \end{cases} \tag{4.3}$$

Thus, summing over c,

$$\Pr\{S_t=X_{t+k}\} = \int_0^1 \frac{1+2u(1-u)}{1-u^2(1-u)^2}(u(1-u))^{|k|}du - \delta_{1k} \tag{4.4}$$

$$= \{\ldots,.001,.002,.011,.050,.244,.383,.244,.050,\ldots\}$$

and, integrating out x, if F is absolutely continuous,

$$\Pr\{C_t=c\} = 4\left(\binom{2c}{c}^{-1} - \binom{2c+2}{c+1}^{-1}\right) - \delta_{1c} \tag{4.5}$$

$$= \{.333,.467,.143,.040,.012,.003,.001,\ldots\} .$$

5. An additive decomposition

In [3] it is shown that if S satisfies (i), (ii), (iii) of section 2 above, and also (v) is of finite span (i.e. $S(\underline{X})_t$ depends only on X_j for $t-a \le j \le t+b$, for some a,b (finite)), and (vi) $\mathrm{Var}(S(\underline{X})_t) \to \infty$; and if furthermore $\underline{X}$ can be written in the form

$$X_t = \mu + Y_t + Z_t \tag{5.1}$$

where $\underline{Y}$ is a stationary Gaussian process, not completely deterministic, and $\underline{Z}$ is a sequence of independent and identically distributed random variables, then

$$S(\underline{X})_t = \mu' + S_L(\underline{Y})_t + R_t$$

where S_L is a linear filter (depending in general on the specification) and $\underline{R}$ is completely orthogonal to $\underline{Y}$ (i.e. $E(R_sY_t) = 0$ for all s,t). We call the term $S_L(\underline{Y})$ the "linear component" of $S(\underline{X})$. The coefficients $\{s_j\}$ of S_L are given by

$$\sum_j E(Y_iY_j)s_{-j} = E(Y_iS(X)_0) \qquad i = 0,\pm1,\pm2,\ldots .$$

or, under a very weak regularity condition, by

$$s_{-j} = E(\partial S(X)_0/\partial X_j) \qquad j = 0,\pm1,\pm2,\ldots .$$

Further, if S is a "selector", i.e. if for almost all t $S(\underline{X})_t = X_{t+k}$ for some k (k may depend on $\underline{X}$), then

$$s_{-j} = \Pr\{S(\underline{X})_t = X_{t+j}\} .$$

The theory developed in [3] does not apply immediately to the smoother 3R, since in [3] it is assumed that S has finite span; however under the specification (5.1) we have $\Pr\{X_{t-1}<X_t<X_{t+1}\} > 0$ so that the r.v. C_t defined before (4.3) is finite wp. 1, and the theory is easily extended to cover the present case.

Thus if $\underline{Y}$ is a process of <u>independent</u> Gaussian variables, with variance σ^2, with component $(3R)_L$ of 3R are given by (4.4) above.

(They do not depend on the distribution of Z_t). Also, from (4.2), for this case we have

$$E(3R(\underline{X})_t)^2 = \frac{2}{3}E\left(X_{2:3}^2\right) + \frac{7}{10}E\left(X_{3:5}^2\right) + \ldots .$$

When $\underline{Z} = 0$, using Teichroew's tables [6] this gives

$$E(3R(\underline{Y})_t^2) = 0.3855\ \sigma^2$$

while from (4.4)

$$E((3R)_L(\underline{Y})_t^2) = 0.2715\ \sigma^2$$

so that in this case

$$E(R_t^2) = 0.1140\ \sigma^2 .$$

An index of the non-linearity of the smoother 3R in this case is thus

$$E(R_t^2)/E(3R(\underline{Y})_t^2) = 0.2957.$$

For the simple 3-median smoother the corresponding calculation gives the value 0.2571.

6. The breakdown probability of a smoother

An important concept in the theory of robust estimation is the 'breakdown point' of an estimator, introduced by Hampel in [2]. In [3] this concept was extended to the smoothing context as follows. Assume $\underline{X} = \underline{Y}+\underline{Z}$ where $\underline{Y}$ is stationary and independent of $\underline{Z}$, which is is a sequence of i.i.d. variables with $\Pr\{Z_t=0\} = p$, but with $\Pr\{Z<z\} = H(z)$ otherwise arbitrary. Then the breakdown probability $B(p)$ of a smoother S is defined to be $\lim_{k\to\infty} \sup_H \Pr\{|S(X)_t|>k\}$. In very many cases (perhaps in all!) B(p) is a function of p but does not depend on the specification of $\underline{Y}$. In crude terms, B(p) is the frequency with which outliers will appear in the output of S when (independent) outliers are present in the input with frequency p. In [3] B(p) is evaluated for several simple smoothers; for smoothers composed of concatenated odd medians it was found that if the longest component median has length 2m+1, then B(p) has the form $ap^{m+1} + O(p^{m+1})$.

Taking $\underline{Y}$ to be an independent process, from the result (4.1) above we find that for the 3R smoother

$$B(p) = \frac{2p^2-p^3}{1-p+p^2} .$$

7. An index of responsiveness

In [4] a simple index of responsiveness to a change in level was defined as

$$J = \sum_{-\infty}^{\infty} \{E(S(\underline{Y}+\underline{d})_t - d_t)^2 - E(S(\underline{Y})_t^2)$$

where Y is an i.i.d. sequence of Normal $(0,\sigma^2)$ variables, and $d_t = 0$, Δ for $t<0$, ≥ 0. In general J depends on both σ^2 and Δ; however if Δ/σ is very large, J is essentially proportional either to Δ^2 or to σ^2, the latter case holding for all concatenated-odd-median smoothers, including 3R. A straightforward extension of the argument in the Appendix shows that when Δ/σ is large, with S = 3R,

$$\Pr\left\{S_{-t} \leq x\right\} = \sum_{j=1}^{t-1} d_j F_{j+1:2j+1}(x) = \left(1 - \sum_{j=1}^{t-1} d_j\right) F_{j+1,2j}(x)$$

(the notation is the same as in (4.2)), so that Teichroew's tables can be used to evaluate J. The result appears in Table 2.

Table 2

Values of J for some smoothers

Smoother	J
3	$1.103\sigma^2$
3^2	$1.318\sigma^2$
3R	$1.377\sigma^2$
5	$2.301\sigma^2$
7	$3.527\sigma^2$

8. Some final remarks

Some unpublished Monte Carlo studies suggest that the linear component 3L of 3R responds rather sensitively to departures from the independence specification. One wonders whether it is possible to design a smoother that has a very similar linear component to 3R in the independence case, with lower non-linearity index, lower breakdown probability, less sensitivity to specification changes, and only moderate worsening of the responsiveness index. The author believes that these properties are achieved by a smoother of the form $A(W(\underline{X}))$ where A is a linear filter, and W is a "(2a+1,2b) Winsorizer", defined as follows: let $R_t(j)$ be the rank of X_{t+j} within the set $(X_{-a}, X_{-2}, \ldots, X_a)$,

$-a \leq j \leq a$. If $b+1 \leq R_t(0) \leq 2a+1-b$ we set $S_t = X_t$; if $R_t(0) \leq b (\geq 2a+2-b)$ we set $S_t = X_{t+k}$ where $R_t(k) = b+1(2a+1-b)$. Details of the calculations that have led to this belief will appear elsewhere.

APPENDIX

Derivation of the results (4.1) and (4.3)

We assume that $\{X_t\}$ are independent with $Pr\{X_t \leq x\} = F = 1-G$ for all t, and enumerate disjoint cases that each lead to the event $E = \{3R(X)_0 < x\}$. First, E occurs if X_{-1}, X_0, X_1 are all $\leq x$, which we denote (---), or if $X_{-1}, X_0 \leq x, X_1 > x$, which we denote (--+), or if $X_1 > x$, $X_0, X_1 \leq x$, which we denote (+--). E cannot occur if both X_0 and X_{-1}, or if both X_0 and X_1, are >x (i.e. for the patterns ++-,-++,+++). Thus

$$Pr\{E\} = F^3 + 2F^2G + F^2GP_3 + FG^2P_3'$$

where

$$P_3 = Pr\{E|(1-+-)\} \ , \ P_3' = Pr\{E|(+-+)\} \ .$$

Now considering the possible values of X_{-2} and X_2, it is easy to check that

$$P_3 = F\ P_5 \quad \text{where} \quad P_5 = Pr\{E|(-+-+-)\}$$

since if either $X_{-2} > x$ or $X_{+2} > x$, E cannot occur. Similarly,

$$P_3 = F^2 + 2FG + G^2P_5 \quad \text{where} \quad P_5' = Pr\{E|(+-+-+)\} \ .$$

The same argument can be repeated, expressing P_5 and P_5 in terms of P_7 and P_7', and so on; the conclusion is

$$P_3 = (2F-F^2)/1-F^2G^2 = P_5 = P_7 = \ldots$$

and the result (4.1) follows.

The derivation of (4.3) is more complicated. Now we assume that $Pr\{x < X_0 \leq x+dx\} = F(dx) \to 0$ as $dx \to 0$. We shall be concerned with events such as $\{X_{-3} < x, X_{-2} > x, X_{-1} > x, x < X_0 < x+dx, X_1 < x\}$, which we denote (-++0+). The zero denotes that at epoch t=0, the value of X_t is in (x,x+dx). We distinguish $4\ \infty^2$ disjoint and exhaustive cases, by specifying $s = sgn(X_{-1}-x)$, $t = sgn(X_1-x)$, the smallest index $b \geq 2$ such that $sgn(X_b-x) = sgn(X_{b-1}-x)$, and the smallest index $a \geq 2$ such that $sgn(X_{-a}-x) = sgn(X_{-a+1}-x)$. We use the notation (a,s 0 t,b) to represent the general pattern; for example (4,-0+,5) represents what in the earlier notation is (--+-0+-+--). On applying the 3R algorithm to this pattern of data, one finds successively the rows of the following array:

$$\begin{array}{c} --+-0+-+-- \\ --000+-- \\ -0\ 00- \\ 0 \end{array} \tag{A.1}$$

Here we have used Obs. 2 to terminate the algorithm as soon as possible at each value of t. Thus for this pattern of data, the central value x appears in the output at epochs t = -1,0,1,2, while the algorithm converges after c = 2,1,2,3 steps respectively. We say that (c,t) = (2,-1) (1,0),(2,1),(3,2) are terminal points for this pattern. Table 1 gives details of all the possible patterns, with their terminal points. It also gives a code letter for each pattern, and its probability; for the pattern studied above this is F^6G^3dF since the pattern has 6-'s and 3+'s

Table 1

Terminal points for all possible patterns

Pattern	Code	Probability	Terminal points	
(2a+1,+0+,2b+1)	A	$F^{a+b+2}G^{a+b}dF$	(a+j+1,j-a)	$0\le j\le b$
			and (a+b+1-j,b-a+j)	$1\le j\le a$
(2a+1,-0-,2b+1)	A^-	$F^{a+b}G^{a+b+2}dF$	same as A	
(2a,+0+,2b+1)	B	$F^{a+b}G^{a+b+1}dF$	(b+j+1,b-j)	$a\le j\le a-1$
(2a,-0-,2b+1)	B^-	$F^{a+b+1}G^{a+b}dF$	same as B	
(2a+1,+0+,2b)	B^R	$F^{a+b}G^{a+b+1}dF$	(a+j+1,j-a)	$0\le j\le b-1$
(2a+1,-0-,2b)	B^{-R}	$F^{a+b+1}G^{a+b}dF$	same as B^R	
(2a,+0-,2b+1)	C	$F^{a+b-1}G^{a+b+2}dF$	(a-j,j+1-a)	$0\le j\le a-1$
			and (j+1,j)	$1\le j\le b$
(2a,-0+,2b+1)	C^-	$F^{a+b+2}\ {}^{a+b-1}dF$	same as C	
(2a+1,+0-,2b)	C^R	$F^{a+b+2}G^{a+b-1}dF$	(a+1-j,j-a)	$0\le j\le a$
			and (j+1,j)	$1\le j\le b-1$
(2a+1,-0+,2b)	C^{-R}	$F^{a+b-1}G^{a+b+2}dF$	same as C	
(2a,+0-,2b)	D	$F^{a+b}G^{a+b}dF$	(a-j,j+1-a)	$0\le j\le a$
			and (j+1,j)	$1\le j\le b-1$
(2a,-0+,2b)	D^-	same as D	same as D	
(2a+1,+0-,2b+1)	E	$F^{a+b+1}G^{a+b+1}dF$	(a+1-j,j-a)	$0\le j\le a$
			and (j+1,j)	$1\le j\le a$
(2a+1,-0+,2b+1)	E^-	same as E	same as E	
(2a,+0+,2b)	F	$F^{a+b+2}G^{a+b-2}dF$	none	
(2a,-0-,2b)	F^-	$F^{a+b-2}G^{a+b+2}dF$	none	

It is now a simple matter to identify the patterns that can lead to the result $3R(X)_0\varepsilon(x,x+dx)$. For example, the pattern studied at (A.1) above corresponds to a set of four ways this can happen, namely when one of $(X_{-3},..,X_6)$, $(X_{-4},..,X_5)$, $(X_{-5},..,X_4)$, $(X_{-6},..,X_3)$ exhibits the pattern (4,-0+,5). From Table 1 we find that for (c,t) = (1,0) we must consider patterns of the types C,C^R,D,E (and their negatives), for (c,t) = (t+1,t) (resp. t+1,-t) (t>1) the types are A,B (resp. B^R),

C, C^R, D, E (and their negatives); and for $(c,t) = (|t|+1+2h, t)$ $(h \geq 1)$ the types are A, B, B^R (and their negatives). In each case the appropriate range of values of a and b has to be determined so that (c,t) falls in the set of terminal points. Assembling all the pieces, (4.3) results.

REFERENCES

[1] J. R. Blum, D. L. Hanson and L. H. Koopmans (1963) On the strong law of large numbers for a class of stationary process. Z. Wahrscheinlichkeits theorie 2, 1-11.

[2] F. R. Hampel (1971) A general qualitative definition of robustness Ann. Math. Statist. 42, 1887-1896.

[3] C. L. Mallows (1979) Some theory of non-linear smoothers. Ann. Statist. 7 xxx-xxx.

[4] C. L. Mallows (1979) Resistant smoothing. Proceedings of the International Time Series Meeting, Nottingham UK March 26-30, 1979.

[5] C. J. Mazreliez and R. D. Martin (1977) Robust Bayesian estimation for the linear model and robustifying the Kalman filter. IEEE Trans. Autom. Control AC-20, 107-110.

[6] D. Teichroew (1956) Tables of expected values of order statistics and products of order statistics from samples of size twenty and less from the normal distribution. Ann. Math. Statist. 27, 410-426

[7] J. W. Tukey (1977) Exploratory Data Analysis, Addison-Wesley, Reading, Mass.

[8] P. F. Velleman (1975) Robust Non-Linear Data Smoothing. Unpublished Ph.D. thesis; Princeton University. (T.R. 89, Series 2, Department of Statistics).

BIAS- AND EFFICIENCY-ROBUSTNESS OF GENERAL M-ESTIMATORS FOR REGRESSION WITH RANDOM CARRIERS

Ricardo Maronna, Oscar Bustos and Victor Yohai

Eidgenössische Technische Hochschule Zürich
Universidade Federal de Pernambuco
and
Universidad Nacional de Buenos Aires

1. Introduction

Let (x_i, y_i) be a sequence of independent identically distributed (iid) random variables, where $\underset{\sim}{x}_i \in R^p$ and $y_i \in R$, and let $\underset{\sim}{\theta} \in R^p$ be an unknown vector such that

(1.1) $$y_i = \underset{\sim}{x}_i{}'\underset{\sim}{\theta} + e_i \quad ,$$

where e_i is independent of $\underset{\sim}{x}_i$ and has distribution H. Although the Least Squares Estimator (LSE) of $\underset{\sim}{\theta}$ is optimal when H is normal, it is nonrobust in the sense that arbitrarily small departures of H from normality may cause arbitrarily large asymptotic variances and/or biases of the estimator. A first step towards robustness is given by the "classical M-estimators" defined as solutions $\underset{\sim}{\theta}^*$ of equations of the form

(1.2) $$\sum_{i=1}^{n} \underset{\sim}{x}_i \psi(y_i - \underset{\sim}{x}_i{}'\underset{\sim}{\theta}^*) = \underset{\sim}{0} \; .$$

When H has density h, the Maximum Likelihood Estimator (MLE) is of this form, with $\psi = h'/h$. If the function ψ is bounded and properly chosen, the efficiency of $\underset{\sim}{\theta}^*$ is "robust" in that it does not diminish remarkably under departures from normality. One example is Huber's psi-function (Huber 1964 and 1973) defined as

(1.3) $$\psi_k(t) = \mathrm{sgn}(t) \min(|t|, k).$$

However, the bias of these estimators is not robust, in that if the joint distribution F of $(\underset{\sim}{x},y)$ follows model (1.1) only approximately, then the asymptotic bias of $\underset{\sim}{\theta}^*$ may be arbitrarily large. This can be measured through two robustness criteria introduced by Hampel (1971 and 1974): the influence function and the breakdown point. These estimators have an unbounded influence function and a null breakdown point (see section 2).

To obtain a robust bias, several families of estimators have been proposed recently, all of the form

$$(1.4) \qquad \sum_{i=1}^{n} \underset{\sim}{x}_i \, w(\underset{\sim}{x}_i)\, \psi((y_i - \underset{\sim}{x}_i'\underset{\sim}{\theta}^*)/v(\underset{\sim}{x}_i)) = \underset{\sim}{0},$$

where $w(\underset{\sim}{x})$ and $v(\underset{\sim}{x})$ are positive functions defined on R^p, with w tending to zero when $\underset{\sim}{x} \to \infty$. The three families:

$$(1.5) \qquad v(\underset{\sim}{x}) = 1, \quad v(\underset{\sim}{x}) = w(\underset{\sim}{x}), \quad v(\underset{\sim}{x}) = 1/w(\underset{\sim}{x})$$

have been respectively proposed by Mallows (1975), Schweppe, and Hill and Ryan (Hill 1977). A typical choice for w is of the type $w(\underset{\sim}{x}) = w_K(|\underset{\sim}{x}|)$, where $|\underset{\sim}{x}|$ is Euclidean norm, and

$$(1.6) \qquad w_K(t) = \psi_K(t)/t$$

with ψ_k defined in (1.3). The particular case of the Schweppe family with $v(\underset{\sim}{x}) = w(\underset{\sim}{x}) = 1/|\underset{\sim}{x}|$ and $\psi = \psi_k$ has been considered by Krasker (1978).

This paper deals with general M-estimators of the form

$$(1.7) \qquad \sum_{i=1}^{n} \underset{\sim}{x}_i \, \varphi(\underset{\sim}{x}_i, r_i) = \underset{\sim}{0} \quad \text{with } r_i = y_i - \underset{\sim}{x}_i'\underset{\sim}{\theta}^*$$

where $\varphi: R^{p+1} \to R$, which includes all former cases. Actually one must normalize the residuals r_i by a scale factor σ^* estimated simultaneously, and the $\underset{\sim}{x}_i$'s by a scatter matrix $\underset{\sim}{S}^*$, i.e., in general one must handle estimators $(\underset{\sim}{\theta}^*, \sigma^*)$ defined as solutions of

$$(1.8) \qquad \sum_{i=1}^{n} \underset{\sim}{x}_i \, \varphi(\underset{\sim}{S}^{*-1}\underset{\sim}{x}_i, r_i/\sigma^*) = \underset{\sim}{0},$$

$$(1.9) \qquad \sum_{i=1}^{n} \chi(|r_i|/\sigma^*) = 0,$$

where $\underset{\sim}{S}^* = \underset{\sim}{S}^*(\underset{\sim}{x}_1,\ldots,\underset{\sim}{x}_n)$ and χ is a monotone function, e.g. the families: $\chi(t) = \psi_k(t)^2 - \beta$ ("Huber's (1964) Proposal 2") or $\chi(t) = \mathrm{sgn}(t-c)$ (Hampel's (1971) normalized median deviation), where β and c are given constants. In (Maronna-Yohai 1978) the consistency and asymptotic normality of $(\underset{\sim}{\theta}^*,\sigma^*)$ are proved under very general assumptions on φ, χ, $\underset{\sim}{S}^*$, and the underlying distribution F, without assuming model (1.1) (in that paper, $\underset{\sim}{S}^*$ appears instead of $\underset{\sim}{S}^{*-1}$). However, the majority of the results of this paper will deal with (1.7), i.e., we shall assume $\underset{\sim}{S}^*$ and σ^* as known constants.

Contents of the paper: In section 2 we calculate the influence function and the breakdown point of estimators (1.7), showing that in order to achieve bias-robustness it is necessary and sufficient that the funtion $\underset{\sim}{x}\ \varphi(\underset{\sim}{x},r)$ be bounded. In section 3 we calculate their efficiency, and show that a bias-robust estimator cannot be efficiency-robust under model (1.1) when $\underset{\sim}{x}$ has heavy tails. In section 4 we compare the robustness and efficiency properties of two families of estimators, and section 5 shows the results of a simulation for finite samples. Section 6 and 7 consider the use of "robust covariances" in regression: either to estimate the matrix $\underset{\sim}{S}^*$ in (1.8) or to estimate $\underset{\sim}{S}^*$, $\underset{\sim}{\theta}^*$ and σ^* simultaneously. Finally, section 8 contains proofs and technical derivations from the preceeding sections.

2. Robustness

Let P be a generic distribution on R^{p+1}, E_P the respective expectation functional, and define $\underset{\sim}{\theta}(P)$ as the solution (if it is unique) of

$$(2.1) \qquad E_P\ \underset{\sim}{x}\ \varphi(\underset{\sim}{x},y-\underset{\sim}{x}'\underset{\sim}{\theta}) = \underset{\sim}{0}\ ,$$

where $(\underset{\sim}{x},y)$ has distribution P. If P_n is the empirical distribution corresponding to a sample of size n, then $\underset{\sim}{\theta}(P_n)$ is equal to $\underset{\sim}{\theta}^*$ defined in (1.7). Under general assumptions (Maronna-Yohai 1978, Huber 1967) it may be proved that $\underset{\sim}{\theta}(P_n) \to \underset{\sim}{\theta}(P)$ when $n \to \infty$. If $\underset{\sim}{\theta}(P)$ is well defined for all P of the form $P = (1-\varepsilon)F + \varepsilon\,\delta(\underset{\sim}{x}_0, y_0)$ - where δ is a point mass - then the influence function of $\underset{\sim}{\theta}^*$ at the distribution F and the point $(\underset{\sim}{x}_0, y_0)$ is (Hampel 1974)

$$(2.2) \qquad IF_\theta(\underset{\sim}{x}_0, y_0) = A^{-1} x_0\, \varphi\,(\underset{\sim}{x}_0, y_0 - \underset{\sim}{x}_0{}'\underset{\sim}{\theta}(F)) ,$$

where

$$(2.3) \qquad \underset{\sim}{A} = \underset{\sim}{A}(F) = E_F\, \varphi'(\underset{\sim}{x}, y - \underset{\sim}{x}'\underset{\sim}{\theta}(F))\ \underset{\sim}{x}\underset{\sim}{x}',$$

where $\varphi'(\underset{\sim}{x}, r) = \partial\varphi(\underset{\sim}{x}, r)/\partial r$. This function gives a good description of the behavior of the asymptotic bias under small perturbations of the model. Clearly, a bounded influence function is equivalent to

$$(2.4) \qquad \sup\{|\underset{\sim}{x}\, \varphi(\underset{\sim}{x}, r)| \ :\ \underset{\sim}{x} \in R^p,\ r \in R\} < \infty.$$

For estimators (1.4) to satisfy (2.4), the functions $\psi(r)$ and $\underset{\sim}{x}\, w(\underset{\sim}{x})$ must be bounded. Clearly, classical M-estimators (1.2) have unbounded influence function.

Define the "gross error sensitivity" as

$$(2.5) \qquad g(\underset{\sim}{\theta}^*) = \sup\{|IF_0(\underset{\sim}{x}_0, y_0)| \ :\ (\underset{\sim}{x}_0, y_0) \in R^{p+1}\}.$$

When H in model (1.1) is normal and $\underset{\sim}{x}$ has a symmetric distribution, it follows from (Hampel 1979) that estimators of the form

$$(2.6) \qquad \varphi(\underset{\sim}{x}, r) \ = \ \psi_k(|\underset{\sim}{x}|r)/|\underset{\sim}{x}|$$

minimize the asymptotic variance of $\underset{\sim}{\theta}^*$ when $p = 1$, under a given bound on the $g(\underset{\sim}{\theta}^*)$. This result was extended by Krasker (1978) by showing that, for general p and $\underset{\sim}{x}$, estimators of this form - after a certain linear transformation of the $\underset{\sim}{x}$'s - minimize the trace of the asymptotic covariance matrix. We shall call them "Hampel-Krasker" (HK) estimators.

We now turn to the breakdown point. We cannot calculate the breakdown point δ^* as defined in (Hampel 1971) because of the complications of Prokhorov distances, but we shall calculate a similar measure of bias-robustness under gross departures from the model: the so-called "gross error breakdown point" ε^*. Let $\mathcal{G}$ be a family of ("contaminating") distributions on R^{p+1}, and $\Theta = R^{p+1}$. Assume that $\underset{\sim}{\theta}(P)$ is well defined for P in a neighborhood of F, and define

$$(2.7) \qquad \varepsilon^* = \varepsilon^*(\underset{\sim}{\theta}^*, F, \mathcal{G}) = \sup\{\varepsilon : \text{there exists } K_\varepsilon \text{ compact} \subseteq \Theta \text{ such that } \underset{\sim}{\theta}[(1-\varepsilon)F + \varepsilon G] \in K_\varepsilon \text{ for all } G \in \mathcal{G}\}.$$

The simplest case is when $\mathcal{G}$ is the family $\mathcal{G}_o$ of all point masses $\delta(\underset{\sim}{x}_0, y_0)$ in R^{p+1}.

Before calculating ε^*, note that with each regression problem one may associate a "natural" subset $X \subseteq R^p$, such that for <u>any</u> conceivable distribution - even the contaminating ones $\underset{\sim}{x}$ should be concentrated on X. For example, if the problem is fitting a straight line (i.e., p=2 and $\underset{\sim}{x} = (1,z)'$, where z is a real random variable) the set X is the line $\{(x_1,x_2): x_1=1\}$. In this case, if an observation is a gross error, the vector $\underset{\sim}{x}$ will still be of the form (1,z) unless, of course, the "1" is also read from punched cards, thus being subjet to punching errors. Similarly, for fitting a parabola :
$X = \{(x_1,x_2,x_3): x_1 = 1, \ x_3 = x_2^2\}$.

We shall work with monotone φ. The reasons are discussed in Remark 1 below.

<u>Theorem 1</u>:

Let $\mathcal{G}$ be any family containing $\mathcal{G}_o$, such that for all $\underset{\sim}{\theta} \in R^p$, $|\underset{\sim}{x}\ \varphi(\underset{\sim}{x}, y-\underset{\sim}{x}'\underset{\sim}{\theta})|$ is integrable with respect to F and to all $G \in \mathcal{G}$. Assume that $\varphi(\underset{\sim}{x},r)$ is nondecreasing and even in r. Let $K(\underset{\sim}{x}) = \sup_r |\varphi(\underset{\sim}{x},r)|$. Then:

a) If $\sup_{\underset{\sim}{x} \in X} |x| \, K(\underset{\sim}{x}) = \infty$, then $\varepsilon^*(\underset{\sim}{\theta}^*,F,\mathcal{G}) = 0$.

b) Otherwise, define for $\underset{\sim}{\gamma} \in R^p$: $s(\underset{\sim}{\gamma}) = \sup_{\underset{\sim}{x} \in X} K(\underset{\sim}{x}) \, |\underset{\sim}{x}'\underset{\sim}{\gamma}|$.

Then if $s(\underset{\sim}{\gamma})$ is continuous

$$\varepsilon^*(\underset{\sim}{\theta}^*,F,\mathcal{G}) = \inf_{\underset{\sim}{\gamma}} \left[E_F K(\underset{\sim}{x}) \, |\underset{\sim}{x}'\underset{\sim}{\gamma}| \, / \, (E_F \, K(x) \, |\underset{\sim}{x}'\underset{\sim}{\gamma}| + s(\underset{\sim}{\gamma}))\right]. \tag{2.8}$$

Proof:

See section 8.1.

Remark 1:

The evenness of φ is assumed only to simplify the proof. The theorem does not require that $\underset{\sim}{\theta}(P)$ be uniquely defined. In effect, the proof shows that when $\varepsilon<\varepsilon^*$, all solutions $\underset{\sim}{\theta}((1-\varepsilon)F+\varepsilon G)$ remain bounded, and that when $\varepsilon > \varepsilon^*$, all tend to infinity.

The situation is not so simple when φ is not monotone, especially when it is "redescending", i.e., $\lim_{r\to\infty} \varphi(\underset{\sim}{x},r)=0$. In this case the different solutions of (2.1) may have very different behaviors under perturbations of F: some may remain bounded and others may tend to infinity. For empirical illustrations see (Hill 1977). Since the estimator is not clearly defined by (2.1) alone, to study its behavior one must also specify the iterative procedure used to solve (2.1), as well as the initial estimator used to start the iterations. Yohai and Klein (1977) follow this approach to study the consistency and asymptotic normality of redescending M-estimators for the linear model.

Remark 2:

Note that ε^* depends on φ only through the function K.

Corollary:

Assume that $\underset{\sim}{x}$ has a spherically symmetric distribution, that φ depends on $\underset{\sim}{x}$ only through $|\underset{\sim}{x}|$, and that $X = R^p$.

Then

a) $\varepsilon^* \leq \varepsilon_p^* = c_p/(c_p+1)$, where $c_p = E\,|z_1|$, where the random vector $\underset{\sim}{z} = (z_1, \ldots, z_p)'$ is uniformly distributed on the spherical surface $\{|\underset{\sim}{z}| = 1\}$.

b) The upper bound ε_p^* is attained when $K(\underset{\sim}{x})\ |\underset{\sim}{x}|$ is constant.

Proof:

See section 8.1.

It is not difficult to obtain an explicit expression for c_p, and from it a recursive formula:

(2.9) $$c_p = (2/\pi)/((p-1)\ c_{p-1}) \text{ with } c_1=1.$$

The values of ε_p^* for p=1,...,5 are respectively: 0.5, 0.39, 0.33, 0,30, 0.27.

Estimators of the form (1.4) attain ε_p^* if ψ is bounded and $w(\underset{\sim}{x}) = 1/|\underset{\sim}{x}|$. In particular, the Mallows estimator with that w, and the Hampel-Krasker estimator, which is seen to optimize simultaneously two criteria.

3. Efficiency

Using the results of (Maronna-Yohai 1978) or (Huber 1967), it may be proved under general regularity assumptions, that the $\underset{\sim}{\theta}^*$ in (1.7) has an asymptotically normal distribution with covariance matrix

(3.1) $$\underset{\sim}{C} = \underset{\sim}{A}^{-1}\ \underset{\sim}{B}\ \underset{\sim}{A}^{-1},$$

where $\underset{\sim}{A}$ is defined in (2.3), and

(3.2) $$\underset{\sim}{B} = \underset{\sim}{B}(F) = E_F\ \varphi(\underset{\sim}{x},\ y-\underset{\sim}{x}'\underset{\sim}{\theta}(F))^2\ \underset{\sim}{x}\underset{\sim}{x}'.$$

In the location case, the introduction of M-estimators with bounded ψ solves simultaneously the problems of obtaining robust efficiency and robust bias. For the general regression case, there exists in the literature an intuitive notion that both are incompatible, see e.g. (Hill 1977) and (Hampel 1977). Intuitively speaking, an observation $(\underset{\sim}{x},y)$, with large $\underset{\sim}{x}$ may have a large influence on the estimator. To prevent the effects of "bad" observations, one must downweight such $\underset{\sim}{x}$. But if the observation is "good", then it contributes much to the precision of $\underset{\sim}{\theta}^*$, and hence should not be downweighted. This is the iron dilemma. We now express it in more formal terms. It has been shown in section 2 that (2.4) is equivalent to bias robustness, as measured by the influence function and the breakdown point. Now we show for the fitting of a straight line through the origin (p=1) that if $\underset{\sim}{\theta}^*$ satisfies (2.4), then its efficiency under model (1.1) may be arbitrarily low when $\underset{\sim}{x}$ is allowed to have heavy tails, in an arbitrarily small neighborhood of the normal F. Let H be a fixed distribution in R with a density h such that $\int|h'(r)|\,dr < \infty$ (e.g., normal). Let G_0 be a fixed distribution on R (e.g., normal), ε_0 a fixed positive number, and $\mathfrak{F}$ the family of all joint distributions $F = F(\varepsilon,G)$ of (x,y), such that under $F : y = x\theta + e$, where x and e are independent, e has density h, and x has distribution $G = (1-\varepsilon)G_0 + \varepsilon G_1$, with G_1 arbitrary and $\varepsilon < \varepsilon_0$. Put $F_0 = F(0,G)$.

Let θ^* be the estimator corresponding to φ, and θ_M any classical M-estimator of the form (1.2). Let $v(\theta^*,F)$ and $v(\theta_M,F)$ be their respective asymptotic variances under F. Thus the relative efficiency under F of θ^* with respect to θ_M is $\mathrm{eff}(F) = v(\theta_M,F)/v(\theta^*,F)$. Then we have:

Theorem 2:
If φ satisfies (2.4), and $\int|h'(r)|\,dr < \infty$, then $\inf_{F\in\mathfrak{F}} \mathrm{eff}(F) = 0$.

Proof:
See section 8.2.

Note that the MLE is the same for all F, namely, as in (1.2) with $\psi = h'/h$. Thus the theorem is valid <u>a fortiori</u> for the absolute efficiency.

Remark:

It may be seen from the proof that the efficiency becomes arbitrarily low by taking ε "small" and G heavy tailed, i.e. "a small proportion of x's widely separated from the rest".

All data analysis for regression is based on residuals, and thus one needs reliable residuals - and hence a reliable $\underset{\sim}{\theta}*$ - to start with, where a "reliable $\underset{\sim}{\theta}*$" means one describing the bulk of the data. This means that - at least from the viewpoint of data analysis - the bias-robust estimators should be preferred. If an observation with large $\underset{\sim}{x}$ is coherent with the rest (i.e., its residual is not large) it represents no problem; otherwise, it may be an erroneous observation, or it may indicate that the model is erroneous (e.g., a quadratic term is needed, or the error variance is not constant). But the decision cannot be reached solely on mathematical grounds. Practical examples of how to handle such situations are shown in (Daniel and Wood, 1971).

In the rest of the paper we concentrate the research on bias-robust estimators.

4. Comparing Mallows and HK estimators

We shall compare the asymptotic efficiency of Mallows (henceforth "MA") and HK estimators for the simplest case: a straight line through the origin. The first one is chosen because of its simplicity, and the second one because of its optimality properties as shown in section 2. The estimators are of the form (1.8)-(1.9). For the first one: $\varphi(x,r) = w_K(|x|) \, \psi_k(r)$, with $k = 1.6$ and $K = 1.73$; for the second one: $\varphi(x,r) = \psi_k(|x|r)/|x|$ with $k = 2.56$. For both estimators, $S* = \text{med} \{|x_i|\} /0.675$, and $X(t) = \text{sgn}(t-0.675)$. The model is: $y = x\theta + e$, with e independent of x; we can take $\theta = 0$. Let $N_0 = N(0,1)$ (standard normal), and let N_1, N_2 and N_3 be contaminated normals of the form $(1-\varepsilon)N(0,1) + \varepsilon N(0,q^2)$, with (ε,q) respectively equal to (0.10; 3), (0.10; 5) and (0.05; 10). The joint density of (x,y) is $f(x,y) = g(x)\,h(y)$, where both g and h are each of N_0, ..., N_3, so that there are 16 situations. The asymptotic values of σ^* for each of the four distributions are respectively: 1.00 , 1.08, 1.10 and 1.06. The parameters of the estimators were chosen so that when $g=h=N_0$, both have asymptotic variance = 1.05. This calibration was necessary to be able to compare many different situations with equanimity.

It follows from (3.1) and the definitions of estimators and distributions, that when $g=N_i$ and $h=N_j$ $(i,j,=0, \ldots, 3)$, the asymptotic variance of MA decomposes as

(4.1) $$\operatorname{var}(MA;\, N_i, N_j) = VX_i VY_j ,$$

where VX_i depends only on w and N_i, and VY_j only on ψ and N_j (it is the asymptotic variance of the location estimator corresponding to ψ). Hence, instead of 16 variances, we can describe MA by giving the VX_i's and VY_j's. When $w(x) \equiv 1$ (i.e., w_K with $K = \infty$) we obtain a classical M-estimator, for which the factor VX is, for $g=N_i$: $VX_i' = (Ex^2)^{-1}$. The quotient $Q_i = VX_i/VX_i'$ indicates the loss of efficiency incurred in when downweighting the x's instead of not doing so. Table 1 gives the values of VX, VY and Q (all numbers in this section are rounded to two decimals). The clearest result is the loss of efficiency for heavy-tailed x, as predicted by Theorem 2.

TABLE 1:

Factors for Mallows estimator

	VY	VX	Q
N_0	1.03	1.02	1.02
N_1	1.30	0.65	1.17
N_2	1.40	0.46	1.48
N_3	1.23	0.37	2.21

Instead of giving the asymptotic variances of HK, we show in Table 2 its deficiencies with respect to MA: var(HK)/var(MA), which are easier to analyse. One sees no spectacular differences between the estimators.

TABLE 2:
Relative deficiency of HK estimator

	h = N0	N1	N2	N3
g = N0	1.00	1.07	1.13	1.13
N1	0.98	1.02	1.08	1.07
N2	0.98	0.98	1.03	1.04
N3	0.94	0.94	0.98	1.00

Although the factorization (4.1) does not hold exactly for the variances of HK, an analysis of the geometric means of the rows and columns of the matrix $\{\mathrm{var}(HK; N_i, N_j)\}$ (i,j=0,...,3) shows that such a factorization holds at least approximately. In table 3 we give factors VX_i, VY_j such that $\mathrm{var}\ (HK;\ N_i, N_j) \cong VX_i VY_j$, where "≅" means "with relative error less than 3%". The factors are chosen so as to coincide with those of MA for the situations i=j=0.

TABLE 3:
Approximate factorization of var(HK)

	VX	VY
N0	1.02	1.03
N1	0.66	1.35
N2	0.40	1.54
N3	0.33	1.33

Table 2, and the comparison between Tables 1 and 3, show that HK is slightly more sensitive than MA to increasing heavy-tailedness of y, and that the reverse happens with x.

In table 4 we give the breakdown point ε^* and gross error sensitivity GES of both estimators for the situations with $h = N_0$.

TABLE 4:
Robustness measures for $h=N_0$

		HK	MA
ε^*	for g = N0	0.50	0.31
GES	for g = N0	2.94	3.40
	N1	2.22	2.66
	N2	1.73	2.15
	N3	1.64	2.07

While the variances of both estimators were not very different, the optimal robustness properties of HK appear more clearly here.

It must be observed that in (Hill, 1977), the same value of k (namely, 1.35) is used for both estimators. This is unfavorable to HK, since it becomes thus less efficient for normal y and for heavy-tailed x.

5. Monte Carlo results

An exploratory simulation was performed, whose main goals were: measuring the loss of efficiency introduced by downweighting large x's, and exploring whether asymptotic theory gives a good information about finite-sample behavior.

The model chosen was (1.1) with p=1, and where each of x and y may be distributed as N_0 or N_2 as previously defined in section 4 (i.e., 4 situations). We used the Mallows estimator: $\varphi = w_K(x)\ \psi_k(y)$, with $k = 1.35$ or $= \infty$, and $K = 1.73$ or $= \infty$, i.e., four estimators. The scale σ^* was simultaneously estimated through "Proposal 2" (see section 1) with $\beta = E_{N_0}\ \psi_k^{\ 2}$, and the scale s^* of x was the normalized median deviation. The sample size was $n=20$, with $N=500$ replications. The variances and pseudo-variances of $n^{1/2}\theta^*$ were estimated. Let $t(\alpha)$ and $\chi^2(\alpha)$ be respectively the α-quantiles of the distribution of $n\theta^{*2}$ and of the chi-squared distribution with one degree of freedom; then the α-pseudovariance is $pv(\alpha) = t(\alpha)/\chi^2(\alpha)$. The variances were estimated by means of the "swindle" methods (Holland 1973), but the quantiles $t(\alpha)$ are the empirical ones. The bisquare ψ was also used (Beaton and Tukey, 1974), but the results were so similar that it was considered not worthwhile to include them here.

Table 5 contains for each situation, the pseudo-variances for $\alpha = 0.30, 0.50, 0.70, 0.90$ and 0.95, the variance and the asymptotic variance. The relative error of the results, as measured the half length of 90% confidence intervals, is between 8% and 13% for pv(0.90), and for variances it is between 2% and 4% when $g = N_0$, and between 7% and 11% when $g = N_2$.

Perhaps the most striking result is the long-tailedness of the distributions when x is N_2, as described by the increasing pv's, which contrast with the almost constant pv's for normal x. Except for the case $k = \infty$ with y distributed as N_2, it seems that the variance is similar to pv(0.95), while the asymptotic variance is similar to pv(0.50).

In view of the clear nonnormailty of θ^*, it would be important to obtain a better approximation to its distribution, and it would be desirable that Monte Carlo studies were not based solely on variances. The long-tailedness of the distributions make variances not one more difficult to interpret, but also imprecise to estimate.

TABLE 5:

Monte Carlo Variances and Pseudo-Variances

(g,h) =	(N2,N2)		(N2,N0)			(N0,N2)			(N0,N0)		
k =	1.34	1.34	∞	1.34	1.34	∞	1.34	1.34	∞	1.34	1.34
K^2 =	∞	3.	∞	∞	3.	∞	∞	3.	∞	∞	3.
α = 0.30	0.34	0.58	0.22	0.22	0.42	2.07	1.55	1.69	1.10	1.15	1.18
0.50	0.43	0.71	0.29	0.32	0.48	2.44	1.61	1.65	1.05	1.16	1.19
0.70	0.57	0.77	0.38	0.39	0.60	2.77	1.58	1.58	1.16	1.11	1.11
0.90	0.86	1.04	0.54	0.57	0.69	3.41	1.69	1.72	1.21	1.25	1.26
0.95	0.91	1.13	0.67	0.66	0.80	3.80	1.66	1.76	1.17	1.24	1.32
Var =	0.86	0.97	0.52	0.54	0.70	3.52	1.60	1.63	1.09	1.14	1.17
As. Var. =	0.40	0.60	0.29	0.31	0.46	3.40	1.38	1.41	1.00	1.05	1.07

6. Normalizing the x's

In this section we consider the choice of the matrix $\underset{\sim}{S}^*$ in (1.8). We do not look for optimality, but we want to compare two intuitively attractive choices. We shall treat a simple situation, namely: model (1.1), with scale σ known and equal to one, and $\underset{\sim}{x}$ having an ellipsoidal distribution (i.e., for some nonsingular pxp-matrix $\underset{\sim}{M}$, the vector $\underset{\sim}{M}\underset{\sim}{x}$ has a spherically symmetric distribution). It may be assumed that $\underset{\sim}{\theta} = \underset{\sim}{0}$.

Let φ depend on $\underset{\sim}{x}$ only through $|\underset{\sim}{x}|$, i.e., $\varphi(\underset{\sim}{x},r)=\varphi_0(\underset{\sim}{x}'\underset{\sim}{x},r)$, where $\varphi_0:R^2 \to R$. Thus the estimator depends on $\underset{\sim}{S}^*$ only through the matrix $\underset{\sim}{V}^* = \underset{\sim}{S}^*\underset{\sim}{S}^{*\prime}$. The simplest way to normalize $\underset{\sim}{x}$ is to divide each coordinate x_j by a scale estimator s_j^*. This amounts to choosing for $\underset{\sim}{V}^*$ a diagonal matrix, say $\underset{\sim}{V}_1^*$. Call $\underset{\sim}{\theta}_1^*$ the respective estimator. Another sound idea is first to transform $\underset{\sim}{x}$ linearly so that it becomes spherically symmetric, and then apply the estimator without normalizing. This amounts to using for $\underset{\sim}{V}^*$ an invariant scatter matrix $\underset{\sim}{V}_2^*$ i.e., if the $\underset{\sim}{x}$'s are transformed to $\underset{\sim}{M}\,\underset{\sim}{x}$, then $\underset{\sim}{V}_2^*$ becomes $\underset{\sim}{M}\underset{\sim}{V}_2^*\,\underset{\sim}{M}'$ for any nonsingular matrix $\underset{\sim}{M}$. This yields an estimator $\underset{\sim}{\theta}_2^*$ invariant under nonsingular linear transformations of $\underset{\sim}{x}$ (i.e., if $\underset{\sim}{x}$ is transformed to $\underset{\sim}{M}\,x$, then $\underset{\sim}{\theta}_2^*$ becomes $\underset{\sim}{M}'^{-1}\underset{\sim}{\theta}_2^*$).

We shall compare $\underset{\sim}{\theta}_1^*$ and $\underset{\sim}{\theta}_2^*$ with respect to robustness (as measured by the gross error sensitivity) and asymptotic variance. We shall work with the Mallows estimator: $\varphi_0(t^2,r) = w_K(t)\,\psi(r)$ where ψ is bounded. It will be assumed that $\underset{\sim}{x}$ is normal, with covariance matrix $\underset{\sim}{U}$, and that for i=1,2 the matrices $\underset{\sim}{V}_i^*$ are consistent, converging to matrices $\underset{\sim}{V}_i$ (depending on $\underset{\sim}{U}$) so calibrated that $\underset{\sim}{V}_i = \underset{\sim}{I}$ when $\underset{\sim}{U}=\underset{\sim}{I}$.

Since both estimators are invariant under coordinatewise linear transformations of $\underset{\sim}{x}$, it may be assumed that the diagonal of $\underset{\sim}{U}$ is composed of ones, i.e.: $\mathrm{var}(x_j)=1$, $j=1,\ldots,p$. Then $\underset{\sim}{V}_1 = \underset{\sim}{I}$ and $\underset{\sim}{V}_2 = \underset{\sim}{U}$ for all $\underset{\sim}{U}$.

Let for i = 1,2

(6.1) $\underset{\sim}{A}_i = E\,\varphi_0'(\underset{\sim}{x}'\underset{\sim}{V}_i^{-1}\underset{\sim}{x},y)\,\underset{\sim}{x}\underset{\sim}{x}'\ ,\ \underset{\sim}{B}_i = E\varphi_0(\underset{\sim}{x}'\underset{\sim}{V}_i^{-1}\underset{\sim}{x},y)^2\,\underset{\sim}{x}\underset{\sim}{x}'$,

where $\varphi_0' = \partial\varphi_0/\partial r$. Then it follows from Lemma 5.2 in (Maronna and Yohai 1978) that under general regularity assumptions, the asymptotic covariance matrix of $\underset{\sim}{\theta}_i{}^*$ is $\underset{\sim}{C}_i = \underset{\sim}{A}_i^{-1}\underset{\sim}{B}_i\underset{\sim}{A}_i^{-1}$. The influence function of $\underset{\sim}{\theta}_i{}^*$ is

(6.2) $\mathrm{IF}_i(\underset{\sim}{x}_0,y_0) = \underset{\sim}{A}_i^{-1}\,\varphi_0(\underset{\sim}{x}_0'\,\underset{\sim}{V}_i^{-1}\,\underset{\sim}{x}_0,y_0)\underset{\sim}{x}_0$

(see section 8.3 for details).

Fortunately, it turns out that the matrices $\underset{\sim}{A}_i$ and $\underset{\sim}{B}_i$ - and hence also $\underset{\sim}{C}_i$ - have the same eigenvectors as $\underset{\sim}{U}$. Let $\underset{\sim}{e}_1,\ldots,\underset{\sim}{e}_p$ be these eigenvectors, with respective eigenvalues $d_1,\ldots,d_p$. Let $\underset{\sim}{T}$ be the orthogonal matrix having the $\underset{\sim}{e}_j$'s as columns. Let us change the coordinate system in the $\underset{\sim}{x}$- and $\underset{\sim}{\theta}$-spaces, transforming $\underset{\sim}{x}$ to $\underset{\sim}{T}'\underset{\sim}{x}$ and $\underset{\sim}{\theta}$ to $\underset{\sim}{T}'\underset{\sim}{\theta}$. In the <u>new</u> coordinates, we have: $\underset{\sim}{U} = \mathrm{diag}\,\{d_1,\ldots,d_p\}$, $A_i = \mathrm{diag}\{a_{i1},\ldots,a_{ip}\}$ and $C_i = \mathrm{diag}\,\{c_{i1},\ldots,c_{ip}\}$, and the coordinates of the influence functions are (putting $\underset{\sim}{x}_0 = (x_{01},\ldots,x_{0p})'$)

(6.3) $\mathrm{IF}_{ij}(\underset{\sim}{x}_0,y_0) = \varphi_0(\underset{\sim}{x}_0'\underset{\sim}{V}_i^{-1}x_0,y_0)/a_{ij}\,x_{0j}\qquad (j = 1,\ldots,p)$

Put $g_{ij} = \sup\{|\mathrm{IF}_{ij}(\underset{\sim}{x}_0,y_0)| : \underset{\sim}{x}_0,y_0\}$. We shall compare variances and gross error sensitivities coordinatewise in the new coordinate system, which seems more appropriate than the original one. It is shown in section 8.3 that

(6.4) $g_{2j} = \beta/(\alpha_K\,d_j^{1/2})\ ,\quad g_{1j} \geq \beta/d_j\ ,$

(6.5) $c_{2j} = \delta\gamma_K/d_j\ ,\quad c_{1j} \geq \delta/d_j\ ,$

where

(6.6) $\beta = K\sup|\psi|/E\psi'(y)\ ,\quad \delta = E\,\psi(y)^2/(E\,\psi'(y))^2,$

(6.7) $\alpha_K = p^{-1}\,E\,w_K(t)\,t^2,\quad \gamma_K = p^{-1}\,E\,w_K(t)^2\,t^2\,/\,\alpha_K^2\ ,$

where t^2 has a chi-squared distribution with p degrees of freedom. If one chooses K large enough that γ_K is not much larger that one (γ_K is the factor "VX" of (4.1)) then also α_K is not much less than one, as the following table shows for some values of p:

p	2	3	5	10
K^2	4	4	4	8
γ_K	1.02	1.03	1.04	1.02
α_K	0.92	0.88	0.79	0.74

Thus with such choices of K, preferring $\underset{\sim}{\theta}_2{}^*$ to $\underset{\sim}{\theta}_1{}^*$ implies: very little loss in all variances, moderate losses in bias-robustness in the direction of largest eigenvalues, and very large gains in the direction of smallest eigenvalues when correlations are large.

The advantages of $\underset{\sim}{\theta}_2{}^*$ are also supported by Hill's (1977) empirical examples of "highly influential" $\underset{\sim}{x}$-points, which are overlooked by analyzing each coordinate separately, but which may be detected after analyzing the values of $\underset{\sim}{x}_i{}'\underset{\sim}{V}^{-1}\underset{\sim}{x}_i$, where $\underset{\sim}{V}$ is the sample covariance matrix.

If one wants $\underset{\sim}{\theta}_2{}^*$ to have a positive breakdown point, and to be consistent under heavy-tailed $\underset{\sim}{x}$, then $\underset{\sim}{V}_2{}^*$ must be a "robust covariance" as in (Maronna 1976) or more generally (Huber 1977). In fact, Hill's examples also show that some $\underset{\sim}{x}$-outliers may remain concealed if one uses the covariance matrix; only repeated trimming brings them to light. A robust covariance would do the same thing more smoothly.

These matrices should not cause great complications when p is small. However, for large p there may appear two drawbacks. One is that for robust covariances, $\varepsilon^* \leq 1/p$ (see above references). The other is that the numerical computation of $\underset{\sim}{V}^*$ may become expensive.

Another use of robust covariances is outlined in the next section.

7. Robust regression from robust covariances

Let $\underset{\sim}{z}_i$ be the (p+1)-dimensional vector with components $\underset{\sim}{x}_i$ and y_i (i.e., $\underset{\sim}{z}_i' = (\underset{\sim}{x}_i', y_i)$). Let u be a positive nonincreasing function, and define the "robust covariance matrix" $\underset{\sim}{W}^*$ as solution of

$$(7.1) \qquad n^{-1} \sum_{i=1}^{n} u(d_i^2) \underset{\sim}{z}_i \underset{\sim}{z}_i' = \underset{\sim}{W}^* \text{ with } d_i^2 = \underset{\sim}{z}_i' \underset{\sim}{W}^{*-1} \underset{\sim}{z}_i .$$

Decompose $\underset{\sim}{W}^*$ in block form:

$$(7.2) \qquad \underset{\sim}{W}^* = \begin{vmatrix} \underset{\sim}{V}^* & \underset{\sim}{c} \\ \underset{\sim}{c}' & b \end{vmatrix} \quad \text{with } \underset{\sim}{V}^* \in R^{p \times p}, \quad \underset{\sim}{c} \in R^p, \ b \in R$$

and define

$$(7.3) \qquad \underset{\sim}{\theta}^* = \underset{\sim}{V}^{*-1} \underset{\sim}{c}, \quad \sigma^{*2} = b - \underset{\sim}{\theta}^{*\prime} \underset{\sim}{V}^* \underset{\sim}{\theta}^*, \qquad r_i = y_i - \underset{\sim}{x}_i' \underset{\sim}{\theta}^* .$$

These expressions may be interpreted in a more familiar way as follows. Express $\underset{\sim}{W}^*$ in the new coordinates $\underset{\sim}{V}^*$, $\underset{\sim}{\theta}^*$ and σ^*. Then the system (7.1) may be equivalently written as

$$(7.4) \qquad n^{-1} \sum_i w_i \underset{\sim}{x}_i r_i = \underset{\sim}{0}$$

$$(7.5) \qquad n^{-1} \sum_i w_i r_i^2 = \sigma^{*2} ,$$

$$(7.6) \qquad n^{-1} \sum_i w_i \underset{\sim}{x}_i \underset{\sim}{x}_i' = \underset{\sim}{V}^*$$

with

$$(7.7) \qquad w_i = u(d_i^2) \text{ and } d_i^2 = \underset{\sim}{x}_i' \underset{\sim}{V}^{*-1} \underset{\sim}{x}_i + r_i^2 / \sigma^{*2} .$$

Thus $\underset{\sim}{\theta}^*$, σ^{*2} and $\underset{\sim}{V}^*$ may be respectively viewed as weighted versions of: LSE of regression, variance of residuals, and covariance matrix of the $\underset{\sim}{x}$'s, with large r's or $\underset{\sim}{x}$'s receiving smaller weights.

One may alternatively interpret (7.4) - (7.5) as (1.8) - (1.9), where $\varphi(\underset{\sim}{x},r) = u(|\underset{\sim}{x}|^2 + r^2)r$, and $\underset{\sim}{S}^*$ is any square root of $\underset{\sim}{V}^*$ (i.e., $\underset{\sim}{S}^*\underset{\sim}{S}^{*'} = \underset{\sim}{V}^*$). Here the matrix $\underset{\sim}{S}^*$ and the function χ depend on both the $\underset{\sim}{x}$'s and r's, instead of only the $\underset{\sim}{x}$'s or the r's as before.

If the function u is such that u(t)t is bounded (which is necessary for robustness) then $\lim_{r \to \infty} \varphi(\underset{\sim}{x},r) = 0$, i.e., the estimator is a "redescending" one. It was proved in (Maronna 1976, Theorem 1) that if u(t) is nonincreasing and t u(t) is bounded and nondecreasing, then under very general assumptions, equation (7.1) possesses a unique solution, and hence also $\underset{\sim}{\theta}^*$ is uniquely defined. One has then a redescending estimator - with its advantages for heavy-tailed errors - without the drawbacks of multiple solutions.

At the same time, note that the weights in (7.4) are small when $\underset{\sim}{x}_i$ is large, <u>even</u> if r_i is small, contrarily to estimators of the form (2.6). Thus estimators (7.4) - (7.6) may be more inefficient for heavy-tailed $\underset{\sim}{x}$ than those in (2.6).

It is clear that $\underset{\sim}{\theta}^*$ is invariant under general linear transformations of $\underset{\sim}{z}_i$. Hence it is regression invariant, and invariant under linear transformations of the $\underset{\sim}{x}$'s.

If $\underset{\sim}{W}^*$ is consistent, then when $n \to \infty$, θ^* tends to a vector $\underset{\sim}{\theta}(F)$, where F is the joint distribution of $(\underset{\sim}{x},y)$. Under the assumptions in (Maronna 1976) mentioned above, this vector is well-defined as a function of F, and it gives an idea of "what does $\underset{\sim}{\theta}^*$ estimate". Let $\underset{\sim}{x}$,y and e denote generically variables distributed as $\underset{\sim}{x}_i, y_i$ and e_i.

<u>Theorem 3:</u>

(a) If $(\underset{\sim}{x},y)$ follow model (1.1) and e has a symmetric distribution, then $\underset{\sim}{\theta}(F) = \underset{\sim}{\theta}$.

(b) If F is ellipsoidal, then $\underset{\sim}{\theta}(F)$ has the property that, for all $\underset{\sim}{\gamma} \in R^p$: $|y - \underset{\sim}{x}'\underset{\sim}{\theta}(F)| \leq_s |y - \underset{\sim}{x}'\underset{\sim}{\gamma}|$, where " $\leq_s$ " means "stochastically not larger than".

Proof:

See Section 8.4.

Besides of their invariance, an advantage of these estimators is that one may perform with the matrix $\underset{\sim}{W}^*$ the same kind of data analysis as with ordinary covariances, e.g., stepwise regression, C_p-plots (Daniel and Wood 1971), etc.. Besides, they may be simpler to compute than the invariant estimator $\underset{\sim}{\theta}_2{}^*$ of section 6, and have approximately the same ε^*. Of course, for larger p they have the drawbacks inherent to robust covariances.

These estimators are consistent and asymptotically normal whenever $\underset{\sim}{W}^*$ is so. The consistency of $\underset{\sim}{W}^*$ was proved in (Maronna 1976) under rather general assumptions, but it would be useful to prove the asymptotic normality of $\underset{\sim}{\theta}^*$ under assumptions broader than Maronna's.

It is still difficult to compare the influence function and asymptotic variances of $\underset{\sim}{\theta}^*$ with those of (1.8) - (1.9), since the formulas are complicated both analytically and numerically.

A regression estimator based on a type of robust covariance proposed by Gnanadesikan was evaluated via Monte Carlo by Denby and Larsen (1977) and found to have very poor performance; but the robust covariance they used, based on "hard trimming" was itself very poor, as may be inferred from (Yohai and Maronna 1976).

8. Appendix

This section contains proofs and derivations from former sections.

8.1 Proof of Theorem 1:

We first prove part (b). Recall that $\varphi(\underset{\sim}{x},r)$ is monotone and even in r.

i) Let $\varepsilon<\varepsilon^*$, then for some constant c: $|\underset{\sim}{\theta}[(1-\varepsilon)F+\varepsilon G]|\leq c$ for all $G\in\mathcal{G}$. Given $\underset{\sim}{\gamma}\neq\underset{\sim}{0}$, choose a sequence $(\underset{\sim}{x}_N,y_N)$ such that $|\varphi(\underset{\sim}{x}_N,y_N-c|\underset{\sim}{x}_N|)|\underset{\sim}{x}_N'\underset{\sim}{\gamma}|\to s(\underset{\sim}{\gamma})$ when $N\to\infty$. Let $G_N=\delta(\underset{\sim}{x}_N,y_N)$ and $P_N=(1-\varepsilon)F+\varepsilon G_N$. Then $\underset{\sim}{\theta}_N=\underset{\sim}{\theta}(P_N)$ satisfies

$$(8.1)\qquad -\varepsilon\varphi(\underset{\sim}{x}_N,y_N-\underset{\sim}{x}_N'\underset{\sim}{\theta}_N)\,\underset{\sim}{x}_N'\gamma=(1-\varepsilon)\,E_F\,\varphi(\underset{\sim}{x},y-\underset{\sim}{x}'\underset{\sim}{\theta}_N)\,\underset{\sim}{x}'\underset{\sim}{\gamma}\;,$$

and $|\underset{\sim}{\theta}_N|\leq c$. Taking modulus in (8.1) and letting $N\to\infty$, one obtains: $\varepsilon\,s(\underset{\sim}{\gamma})\leq(1-\varepsilon)\,E_F\,K(\underset{\sim}{x})|\underset{\sim}{x}'\underset{\sim}{\gamma}|$, and this implies that ε is not greater than the right-hand side of (2.8).

ii) Let $\varepsilon>\varepsilon^*$. Then there exists a sequence $G_N\in\mathcal{G}$ such that $\underset{\sim}{\theta}_N=\underset{\sim}{\theta}(P_N)$ tends to infinity. Let $\underset{\sim}{\gamma}_N=\underset{\sim}{\theta}_N/|\underset{\sim}{\theta}_N|$. Since $|\underset{\sim}{\gamma}_N|=1$, there exists a convergent subsequence $\underset{\sim}{\gamma}_{N'}$. Let $\underset{\sim}{\gamma}=\lim\underset{\sim}{\gamma}_{N'}$. From (2.1) we have

$$(8.2)\qquad 0=(1-\varepsilon)\,E_F\varphi(\underset{\sim}{x},y-\underset{\sim}{x}'\gamma_{N'}|\underset{\sim}{\theta}_{N'}|)\underset{\sim}{x}'\gamma_{N'}+\varepsilon\,E_{G_{N'}}\varphi(\underset{\sim}{x},y-\underset{\sim}{x}'\underset{\sim}{\theta}_{N'})\underset{\sim}{x}'\gamma_{N'}$$

Since $y-\underset{\sim}{x}'\underset{\sim}{\gamma}_{N'}|\theta_{N'}|\to(-\infty)\,\mathrm{sgn}\,(\underset{\sim}{x}'\underset{\sim}{\gamma})$, (8.2) implies when $N'\to\infty$: $0\leq(1-\varepsilon)E_F\,K(\underset{\sim}{x})\,|\underset{\sim}{x}'\underset{\sim}{\gamma}|+\varepsilon\,s(\underset{\sim}{\gamma})$. This implies that ε is not less than the right-hand side of (2.8).

To prove part (a), assume that $\varepsilon^*>0$, take $0<\varepsilon<\varepsilon^*$, and choose $(\underset{\sim}{x}_N,y_N)$ such that $|\varphi(\underset{\sim}{x}_N,y_N-|\underset{\sim}{x}_N|c)\,\underset{\sim}{x}_N|\to\infty$. Let $\underset{\sim}{\gamma}_N=\underset{\sim}{x}_N/|\underset{\sim}{x}_N|$. By proceeding as in (8.1) with $\underset{\sim}{\gamma}=\underset{\sim}{\gamma}_N$, we obtain a contradiction.

Proof of the Corollary:

Let $\underset{\sim}{z} = \underset{\sim}{x}/|\underset{\sim}{x}| = (z_1,\ldots,z_p)'$. Since $\underset{\sim}{z}$ is independent of $|\underset{\sim}{x}|$, and $K(\underset{\sim}{x})$ depends only on $|\underset{\sim}{x}|$, we have for all $\underset{\sim}{y}$: $E\, K(\underset{\sim}{x})\,|\underset{\sim}{x}'\,\underset{\sim}{y}| = c_p\, E\, K(\underset{\sim}{x})\,|\underset{\sim}{x}|$. Let $e = E\, K(\underset{\sim}{x})\,|\underset{\sim}{x}|$ and $f = \sup_{\underset{\sim}{x}} K(\underset{\sim}{x})\,|\underset{\sim}{x}|$. Then by (2.8), $\varepsilon^* = e/(e+f)$, which is an increasing function of e. Hence the maximum is obtained when $K(\underset{\sim}{x})\,|\underset{\sim}{x}|$ is identically equal to f.

To calculate c_p, let $\underset{\sim}{x}$ be spherical normal, and define $\underset{\sim}{z}$ as above. Then $\underset{\sim}{z}$ is uniformly distributed on $\{|\underset{\sim}{z}| = 1\}$, and z_1^2 has an F distribution with degrees of freedom 1 and p. This yields

$$c_p = 2\pi^{-1/2}\Gamma\,(p/2)/\left[\Gamma((p-1)/2)\;(p-1)\right],$$

and this implies (2.9).

8.2 Proof of Theorem 2:

One may assume $\theta = 0$. To simplify notation, put $\Psi\,(x,r) = x\,\varphi(x,r)$ and $\Psi'(x,r) = x\,\varphi'(x,r)$. The asymptotic variances of θ^* and θ_M under F are respectively

$$v(\theta^*,F) = E_F\,\Psi(x,y)^2/(E_F\,\Psi'(x,y)x)^2\;, \quad v(\theta_M,F) = c/E_G\,x^2,$$

where $c = E_H\,\psi(y)^2\,/\,(E_H\,\psi'(y))^2$.

Let $G_1 = \delta(x_0)$. Then $\mathrm{eff}(F(\varepsilon,G)) = A^2/B$, where

$$A = (1-\varepsilon)\,E_{F_0}\,\Psi'(x,y)\,x + \varepsilon\,E_H\,\Psi'(x_0,y)\,x_0$$

$$B = [(1-\varepsilon)\,E_F\,\Psi(x,y)^2 + \varepsilon E_H\,\Psi(x_0,y)^2][(1-\varepsilon)E_{G_0}\,x^2 + \varepsilon\,x_0^2].$$

Recall that $|\Psi(x,y)| \leq b$ for some constant b, and note that integration by parts yields $|E_H\;\Psi'(x_0,y)| \leq b\int|h'(y)|\,dy$. By dividing A^2 and B by x_0^2 one obtains: $\liminf_{x_0\to\infty} \mathrm{eff}(F(\varepsilon,G)) \leq \varepsilon$ where a is a constant independent of ε. This proves the Theorem.

8.3 Here we give the derivations needed in section 6. Let $\underset{\sim}{\theta}{}^*$ satisfy (1.8) and let $IF_V(\underset{\sim}{x}_0) \in R^{p \times p}$ be the influence function of the matrix $\underset{\sim}{V}{}^*$. Then if $\underset{\sim}{\theta}{}^* \to \underset{\sim}{\theta}(F)$ where F is the joint distribution of $(\underset{\sim}{x}, y)$, and $\underset{\sim}{V}{}^* \to V$ when $n \to \infty$, a straightforward and laborious calculation shows that the influence function of $\underset{\sim}{\theta}{}^*$ is

$$IF_\theta(\underset{\sim}{x}_0, y_0) = \underset{\sim}{A}{}^{-1} \; [\varphi_0(\underset{\sim}{x}_0' \underset{\sim}{V}{}^{-1} \underset{\sim}{x}_0, y_0 - \underset{\sim}{x}_0' \underset{\sim}{\theta}(F)) \, \underset{\sim}{x}_0$$

$$- E_F \dot{\varphi}_0 \; (\underset{\sim}{x}' \underset{\sim}{V}{}^{-1} \underset{\sim}{x}, y - \underset{\sim}{x}' \underset{\sim}{\theta}(F)) \; \underset{\sim}{x} \quad \underset{\sim}{x}' \underset{\sim}{V}{}^{-1} IF_V(\underset{\sim}{x}_0) \underset{\sim}{V}{}^{-1} \underset{\sim}{x}],$$

where $\underset{\sim}{A}$ is as in (6.1), and $\dot{\varphi}_0(t, r) = \partial \varphi_0(t, r) / \partial t$. Assume φ_0 even in r. If the joint distribution of $(-\underset{\sim}{x}, y)$ or of $(\underset{\sim}{x}, -y)$ equals F (as happens with the model in section 6) then the last term vanishes and one is left with an expression like (6.2). We may hence forth assume $\underset{\sim}{\theta}(F) = \underset{\sim}{0}$.

Put now $\underset{\sim}{v} = (v_1, \ldots, v_p)' = \underset{\sim}{T}' \underset{\sim}{x}$, so that the v_j's are uncorrelated normals. Then in the new coordinate system, the matrix $\underset{\sim}{A}_2$ becomes

(8.3) $\qquad A_2 = \underset{\sim}{T}' \underset{\sim}{A}_2 \underset{\sim}{T} = E \, \varphi_0' \; (\underset{\sim}{v}' \underset{\sim}{D}{}^{-1} \underset{\sim}{v}, y) \; \underset{\sim}{v} \underset{\sim}{v}' \, ,$

where $D = \text{diag} (d_1, \ldots, d_p)$. Then the assumed symmetry of F implies that $\underset{\sim}{A}_2$ is diagonal. The same procedure shows that the matrix

(8.4) $\qquad \underset{\sim}{A}_1 = \underset{\sim}{T}' \underset{\sim}{A}_1 \underset{\sim}{T} = E \, \varphi_0' (\underset{\sim}{v}' \underset{\sim}{v}, y) \; \underset{\sim}{v} \underset{\sim}{v}'$

is diagonal. The reasoning for $\underset{\sim}{B}_i$ is the same.

Put now $z_j = v_j / d_j^{1/2}$, so that $\underset{\sim}{z} = (z_1, \ldots, z_p)'$ has identity covariance matrix. The spherical symmetry implies that the elements of the diagonal of $\underset{\sim}{A}_2$ are

$$a_{2j} = p^{-1} E \, \varphi_0' \; (\underset{\sim}{z}' \underset{\sim}{z}, y) \; \underset{\sim}{z}' \underset{\sim}{z} \, d_j \qquad (j=1, \ldots, p)$$

while those of A_1 are

$$a_{1j} = E\varphi_0' \ (\underset{\sim}{z}'\underset{\sim}{D}^{-1}\underset{\sim}{z}, y)\ z_j^2\ d_j . \quad (j=1,\ldots,p)$$

For the Mallows estimator, one obtains easily for $\underset{\sim}{\theta}_2{}^*$: $a_{2j} = (E\,\psi'(y)\,\alpha_K\ d_j$, with α_K as in (6.7). Since the function $t\,w_K(t)$ is nondecreasing and reaches its supremum K when $t \to \infty$, it results

$$\sup_{\underset{\sim}{x}_0} w_K\left[\left(\sum_j x_{0j}{}^2/d_j\right)^{1/2}\right]|x_{0j}| = K/d_j^{1/2}.$$

This proves the first equation in (6.4). As for $\underset{\sim}{\theta}_1{}^*$:

$$\sup_{\underset{\sim}{x}_0} w_K\left[\left(\sum_j x_{0j}{}^2\right)^{1/2}\right]|x_{0j}| = K ,$$

and since $w_K \leq 1$, $E\,w[(\underset{\sim}{z}'\underset{\sim}{D}^{-1}\underset{\sim}{z})^{1/2}]z_j{}^2 \leq E\,z_j{}^2 = 1$, and thus $a_{1j} \leq d_j\,E\psi'(y)$. This proves the second part of (6.4).

A similar reasoning leads to (6.5), where the Cauchy-Schwarz inequality must be used to prove that

$$E\,w_K\left[(\underset{\sim}{z}'\underset{\sim}{D}^{-1}\underset{\sim}{z})^{1/2}\right]^2 z_j{}^2 \;/\; (E\,w_K\left[(\underset{\sim}{z}'\underset{\sim}{D}^{-1}\underset{\sim}{z})^{1/2}\right] z_j{}^2)^2 \geq 1 .$$

The bounds for $\underset{\sim}{\theta}_1{}^*$ given in (6.4) and (6.5) may be sharpened, but the correction is small in relation to the values for $\underset{\sim}{\theta}_2{}^*$.

<u>8.4</u> Now we give the derivations corresponding to section 7. It is easy to derive (7.4) - (7.5) - (7.6) from (7.1). The second equation in (7.7) follows from the easily verifiable decomposition

$$\underset{\sim}{W}^{*-1} = \begin{bmatrix} \underset{\sim}{D} & \underset{\sim}{f} \\ \underset{\sim}{f}' & e \end{bmatrix}$$

with $\underset{\sim}{D} = \underset{\sim}{V}^{-1} + \underset{\sim}{\theta}^*\underset{\sim}{\theta}^{*\prime}/\sigma^{*2}$, $e = 1/\sigma^{*2}$, $\underset{\sim}{f} = -\underset{\sim}{\theta}^*/\sigma^{*2}$.

<u>Proof of Theorem 3:</u>

(a) Since the estimate is regression invariant, one may assume that $\underset{\sim}{\theta} = \underset{\sim}{0}$ in (1.1). The result follows from the symmetry of y.

(b) Since $\underset{\sim}{\theta}(F)$ is invariant under general linear transformations of $\underset{\sim}{z}$, one may assume F to be spherically symmetric, and thus $\underset{\sim}{\theta}(F) = \underset{\sim}{0}$. Let $\underset{\sim}{\beta}' = (\underset{\sim}{\gamma}', -1) \in R^{p+1}$, so that we must show that $|y| \leq_s |\underset{\sim}{z}'\underset{\sim}{\beta}|$.

Let $\underset{\sim}{\alpha}' = (\underset{\sim}{0}' |\underset{\sim}{\beta}|) \in R^{p+1}$, and let $\underset{\sim}{T}$ be an orthogonal transformation in R^{p+1} such that $\underset{\sim}{T}'\underset{\sim}{\beta} = \underset{\sim}{\alpha}$. Note that $\underset{\sim}{T}\underset{\sim}{z} =_s \underset{\sim}{z}$ (where "$=_s$" means "has the same distribution") and hence

$$|\underset{\sim}{z}'\underset{\sim}{\beta}| =_s |(\underset{\sim}{T}\underset{\sim}{z})'\underset{\sim}{\beta}| = |\underset{\sim}{z}'\underset{\sim}{\alpha}| = |\underset{\sim}{\beta}||y| ,$$

and this is $\geq |y|$, since $|\beta|^2 = |\gamma|^2 + 1 \geq 1$.

REFERENCES

Beaton, A.E. and Tukey, J.W. (1974). The fitting of power series, meaning polynomials, illustrated on bandspectroscopic data. Technometrics 16, 147-185.

Daniel, C. and Wood, F.S. (1971). Fitting equations to data. Wiley, New York.

Denby, L. and Larsen, (1977). Robust regression estimators compared via Monte Carlo. Comm. Stat. A6(4), 335-362.

Hampel, F.R. (1971). A general qualitative definition of robustness. Ann. Math. Statist. 42, 1887-1896.

Hampel, F.R. (1974). The influence curve and its role in robust estimation. J. Amer. Statist. Assoc. 69, 383-394.

Hampel, F.R. (1977). Modern trends in the theory of robustness. Research report No. 13. Fachgruppe für Statistik - ETH Zürich.

Hampel, F.R. (1978). Optimally bounding the gross-error-sensitivity and the influence of position in factor space. Research report No. 18. Fachgruppe für Statistik - ETH Zürich.

Hill, R.W. (1977). Robust regression when there are outliers in the carriers. Ph.D.Thesis, Statistics Department, Harvard University.

Holland, P.W. (1973). Monte Carlo for robust regression: the swindle unmasked. Working paper No. 10. National Bureau of Economic Research, Inc.

Huber, P.J. (1964). Robust estimation of a location papremeter. Ann. Math. Statist. 35, 73-101.

Huber, P.J. (1967). The behavior of maximum likelihood estimates under nonstandard conditions. Proc. Fifth Berkeley Symp. Prob. 1, 221-233. University of California Press.

Huber, P.J. (1973). Robust regression: Asymptotic, conjectures, and Monte Carlo. Ann. Statist. 1, 799-821.

Huber, P.J. (1977). Robust covariances. In: Statistical Decision Theory and Related Topics II, 165-191. Academic Press, New York.

Krasker, W.S. (1978). Estimation in linear regression models with disparate data points. Unpublished manuscript.

Mallows, C. (1975). On some topics in robustness. Unpublished memorandum, Bell Telephone Laboratory, Murray Hill.

Maronna, R.A. (1976). Robust M-estimators of multivariate location and scatter. Ann. Statist. 4, 51-67.

Maronna, R.A. and Yohai, V.J. (1978). Asymptotic behavior of general M-estimators for regression and scale with random carriers. Submitted for publication.

Yohai, V.J. and Klein, R. (1978). Asymptotic behavior of iterative M-estimators. Unpublished manuscript.

Yohai, V.J. and Maronna, R.A. (1976). Location estimators based on linear combinations of modified order statistics. Comm. Stat. A5, 481-486.

APPROXIMATE CONDITIONAL-MEAN TYPE SMOOTHERS AND INTERPOLATORS[†]

R. Douglas Martin[††]

Abstract

A class of robust smoother and interpolator algorithms is introduced. The motivation for these smoothers and interpolators is a theorem concerning approximate conditional-mean smoothers for vector Markov processes in additive non-Gaussian noise. This theorem is the smoothing analog of Masreliez's approximate non-Gaussian filter theorem (IEEE-Auto. Control, AC-20, 1975). The theorem presented here relies on the assumption that a certain conditional density is Gaussian, just as does Masreliez's result. This assumption will rarely, if ever, be satisfied exactly. Thus a continuity theorem is also presented which lends support to the intuitive notion that the conditional density in question will be nearly Gaussian in a strong sense when the additive noise is nearly Gaussian in a comparatively weak sense. Approaches for implementing the robust smoothers and interpolators is discussed and an application to a real data set is presented.

[†] Invited talk delivered at the Heidelberg Workshop on Smoothing Techniques for Curve Estimation, Heidelberg, Germany, April 2-4, 1979.

[††] Department of Electrical Engineering, University of Washington, Seattle, Washington 98195; Consultant, Bell Laboratories, Murray Hill, New Jersey 07974.

This research was supported in part by NSF Grant ENG 76-00504.

1. INTRODUCTION

Perhaps the currently best-known type of robust smoothers are those based on moving order statistics as introduced by J. W. Tukey (1977a, 1977b). The simpliest example of such a smoother would be a moving median of prescribed span. It should be noted however that Tukey in fact introduced a large variety of fancier smoothers based on concatenations of moving order statistics and linear smoothers, along with a device called "twicing." By analogy to the use of the term "L-estimate" to describ any of a broad class of location parameter estimates based on order statistics, we shall refer to both plain and fancy smoothers based on moving order statistics as "L-smoothers."

Following Tukey's lead, P. Velleman (1975) studied the behavior of certain L-smoothers via Monte Carlo for a variety of input processes consisting of a purely sinusoidal function and additive noise with and without outliers. A motivation for using the latter as inputs is that the response to purely sinusoidal inputs for all frequencies between zero and one-half provides a complete description of a linear smoother in the form of the "transfer function." Of course this is not the case for robust smoothers which are inherently non-linear. Non-linearity can cause transfers of power from one frequency to another (cf. Velleman, 1975).

Papers on L-smoothers have recently begun to appear in the engineering literatur (Rabiner, Sambur and Schmidt, 1975; Justusson, 1977; Huang, Yang and Tang, 1979). This is no doubt due to the real need for some kind of robust smoothing to deal with outliers in time and space series, along with the fact that L-smoothers have rather obvious and intuitively appealing resistance properties.

If L-smoothers provide worthwhile robust smoothers, then it would come as no surprise to find that moving maximum-likelihood type estimates (M-estimates) of location (P. Huber, 1964) provide useful robust smoothers. One can find pertinent discussions in the recent work of Cleveland (1979) and Stuetzle (1979). Interestingl enough the historical order of development parallels that of location estimates - medians and trimmed means were used by data analysts who were concerned about outlier long before M-estimates came into public view.

Yet another interesting and potentially quite useful class of robust smoothers is that based on robustified splines. P. Huber's (1979) recent paper is a basic reference to this approach. As Huber mentions, a number of others have considered robustifying splines, but little has appeared in the literature so far (see however Lenth, 1977).

The various approaches to robust smoothing just described all share the common virtue of being resistant to outliers. They also have other desirable properties such as translation equivariance, shift equivariance, etc. However, a detailed understanding of their properties in probabilistic terms has been lacking since until recently, there has been a scarcity of tools with which to do a careful statistical analysis of nonlinear smoother behavior. Thus it has been difficult for potential users to determine which of several approaches, and which particular smoother within a given class, will be a good one for his problem.

The above state of affairs may not have been too intolerable in the context of exploratory data analysis where the main use of robust smoothing is to reveal structure in the data which is not easily detected by eye in the raw data. Several striking examples have been given by Tukey (1977a, 1977b). On the other hand there are other important goals of robust smoothing, such as obtaining outlier free time-series for the purpose of fitting autoregressive-moving-average models (Martin, 1979a, 1979b) or estamating spectral density functions (Kleiner, Martin and Thomson, 1979). For such purposes one will naturally require much more detailed knowledge of the smoother's statistical behavior in order to determine the impact of robust smoothing on the final estimates derived from the smoothed series. Such knowledge should also prove quite useful for selecting robust smoothers for use in exploratory data analysis.

Recently Mallows (1979a, 1979b) has made a significant contribution to the theory of nonlinear smoothers which should greatly enhance our ability to analyze proposed robust smoothers of many varieties. A most important aspect of his work is a theorem which characterizes the "linear part" of a nonlinear smoother, and provides an additive orthogonal decomposition of the smoother into the linear part and a residuals process.

Presumably a good robust smoother would have a linear part which is "close" to the linear smoother which the user would prescribe for an outlier-free process (if one could ever be assured of such a nicety in advance), and a residual process which is relatively "small." Mallows (1977a) has estimated the linear-part transfer function of a few L-smoothers via Monte Carlo for a variety of input processes (N.B., the linear-part transfer function of a nonlinear smoother depends upon the probabilistic specification of the input process, unlike as in the case of linear smoothers). More extensive results of this type, when available in published form, may provide useful guidelines for potential users of robust smoothers.

It should be noted that Mallows' decomposition theorem and associated Monte Carlo studies of linear-part transfer functions are primarily of use for the <u>analysis</u> of robust smoothers. Mallows addresses the issue of how to <u>design</u> a robust smoother with prescribed characteristics by only a brief comment to which I shall return later on.

Furthermore, the literature in general reveals no attempts to design robust smoothers which are optimal for particular non-Gaussian process model specifications. This situation seems particularly unsatisfactory in the following light: (i) M-estimates of location and regression parameters are asymptotically efficient maximum-likelihood estimates for appropriate choices of the psi-function ψ which operates on the residuals in the estimating equations, and (ii) P. Huber's (1964) seminal paper teaches us that a robust estimate may be obtained by using the maximum-likelihood estimate for an appropriate heavy-tailed error distribution.

While the maximum-likelihood estimate for the least-favorable distribution in a specified family is asymptotically mini-max (under suitable conditions), we should not regret too much giving up this property providing we can still find a highly robust estimate. Fortunately the use of error distributions which are not least-favorable but which are sufficiently smooth, and have tails of at least exponential thickness, results in a qualitatively robust estimate (cf. Hampel, 1971), at least for location problems. Thus there is considerable motivation to search for the general structure of robust estimates in more complicated problems (where hopes for finding a mini-max estimate appear dim) such as the smoothing problem, by finding the structure of optimal estimates for sufficiently heavy-tailed error distributions in a reasonably realistic model formulation.

This paper is motivated by the above attitude. As a "reasonably realistic" model formulation one might assume that the series of interest is representable as an autoregressive-moving-average (ARMA) process with outliers modeled in an additive form. The basic rationale here is: (i) if a series indeed has smooth structure to be found, and the series is reasonably homogeneous, then the associated correlation structure can often be well approximated by an ARMA model - and often low-order autoregressive (AR) models will do; (ii) a non-Gaussian additive outlier (AO) model will be a reasonable and useful approximation for many, though obviously not all, time series with outliers; (iii) we have much to learn from finding out what a more-or-less good approach to robust smoothing is in this context.

The calculation of an exact optimal smoother (i.e., the conditional-mean, which is optimal using a mean-squared-error criterion) for additive non-Gaussian noise turns out to be rather intractable. Thus an approximate non-Gaussian smoother has been obtained using the assumption that a certain conditional density is Gaussian. This device was used by Masreliez (1975) to derive an approximate conditional-mean filter for non-Gaussian additive noise.† Although the key assumption just cited is not satisfied in most if not all specific non-Gaussian additive noise situations, it appears to be "very nearly" satisfied. Furthermore the resulting approximate conditional-mean smoother has a strong intuitive appeal which echoes that of Masreliez's filter.

The paper is organized in the following way. Section 2 defines the non-Gaussian time-series model used, reviews Masreliez's approximate conditional-mean filter result and provides details concerning the structure and implementation of approximate conditional-mean type filters for the special case of autoregressions. In Section 3 a theorem is given which reveals the structure of approximate conditional-mean (ACM)

† In this paper a "filter" uses observations $y_1, \ldots, y_k$ to produce filtered values $\hat{x}_k$, $1 \le k \le n$ whereas a "smoother" uses all the data $y_1, \ldots, y_n$ to produce smoothed values $\hat{x}_k^n$, $1 \le k \le n-1$. This is in keeping with common engineering terminology, and with this terminology a filter may be regarded as a one-sided smoother.

type smoothers. The proof of this theorem is deferred to Section 4, and Section 3 continues by discussing the implementation of ACM-smoothers for autoregressions and presenting an example of using such a smoother on real data. Section 5 presents a continuity theorem which suggests that the conditional density assumed to be exactly Gaussian in the proof of the ACM-smoother theorem will be nearly Gaussian when the additive noise is nearly Gaussian.

2. APPROXIMATE CONDITIONAL-MEAN TYPE FILTERS

The robust smoothers presented in Section 3 make use of a robust filter as a basic ingredient. The robust filter is an approximate conditional-mean type filter motivated by Masreliez's (1975) result. This section expands upon the rather sketchy treatment of such robust filter in Martin (1979a) by providing important details left out in the cited reference. The treatment given here concentrates on autoregressive models, and also overlaps somewhat with Martin (1979b) where robust filters for moving-average and autoregressive-moving-average models are discussed.

In this section it is assumed that the parameters are known for a good autoregressive approximation to the time series at hand. Estimation of the parameters is discussed in Section 3.

Masreliez's Theorem The formal setup of Masreliez's theorem is a conventional linear-state-variable model often used for estimation and control theory in the engineering literature. Since we are concerned with univariate time series problems, a scalar-observations version of the model is used here. It is supposed that the scalar observations $y_1, \ldots, y_n$ are well modeled as the sum of a location parameter, a scalar noise process v_k and a linear transformation of a first-order vector Markov process as follows:

$$\underline{x}_k = \Phi\, \underline{x}_{k-1} + \underline{\varepsilon}_k$$
$$y_k = \mu + H\underline{x}_k + v_k\,. \tag{2.1}$$

It is assumed that $\underline{x}_k$ and $\underline{\varepsilon}_k$ have dimension p, Φ is a p x p matrix and H is a 1 x p matrix. It is also assumed that $\underline{x}_k$ is independent of future $\underline{\varepsilon}_k$, and that $\underline{\varepsilon}_k$, v_k are mutually independent sequences which are individually independent and identically distributed (i.i.d.). Although stationarity of $\underline{\varepsilon}_k$ and v_k are assumed for purposes of this paper, both the theorem due to Masreliez presented in this section and the smoothing theorem of Section 3 are valid when $\underline{\varepsilon}_k$, v_k, Φ, and H are time-varying. While the time-varying version is important in some engineering applications, it is not needed in the context of modeling stationary time series which is the main interest here.

The pth-order autoregression

$$y_k - \phi_1 y_{k-1} - \ldots - \phi_p y_{k-p} = \varepsilon_k \tag{2.2}$$

with location parameter μ is written in the state-variable from (2.1) by setting

$$\Phi = \begin{pmatrix} \phi_1 & \phi_2 & \cdots & & \phi_p \\ \hline 1 & 0 & \cdots & & 0 \\ 0 & 1 & 0 \cdots & & 0 \\ \cdot & & & & \cdot \\ \vdots & & & & \vdots \\ \cdot & & & 0 & \cdot \\ 0 & 0 & \cdots & 0\ 1 & 0 \end{pmatrix} \tag{2.3}$$

$$\underline{\varepsilon}_k^T = (\varepsilon_k, 0, \ldots, 0)$$

and

$$H^T = (1, 0, \ldots, 0) .$$

It is assumed that y_k is stationary.

The notation to be used throughout is as follows. Denote the first k observatio[ns] by $Y^k = (y_1, y_2, \ldots, y_k)$. The "state" or "signal" prediction density $f(\underline{x}_{k+1}|Y^k)$ is the conditional density of $\underline{x}_{k+1}$ given Y^k. This density is assumed to exist for $k \geq 1$. The observation prediction density is $f(y_{k+1}|Y^k)$. These densities of course depend upon μ, H, Φ and the distributions of $\underline{\varepsilon}_k$, v_k. The conditional-mean estimate of $\underline{x}_k$ given Y^k is written $\hat{\underline{x}}_k = E\{\underline{x}_k|Y^k\}$ and the conditional-mean estimate of $\underline{x}_{k+1}$ given Y^k is written $\hat{\underline{x}}_{k+1}^k = E\{\underline{x}_{k+1}|Y^k\}$. Assuming they exist, these conditional-means are the minimum mean-squared-error estimates given Y^k, under mild distributional assumptions (Meditch, 1969). It is easily checked that $\hat{\underline{x}}_{k+1}^k = \Phi\, \hat{\underline{x}}_k$ under the model assumptions.

When $\underline{\varepsilon}_k$ and v_k are Gaussian the straightforward computation of $\hat{\underline{x}}_k = E\{\underline{x}_k|Y^k\}$ by any one of a variety of techniques (Jazwinski, 1970) yields the Kalman filter recursion equations. Unfortunately, the explicit and exact calculation of $\hat{\underline{x}}_k$ in closed form is virtually intractible when v_k is non-Gaussian (except perhaps in a few special cases, e.g., as in Stuck, 1976). However, there is a simplifying assumption, discovered by C. J. Masreliez (1975), which allows one to make an exact calculation of $\hat{\underline{x}}_k$, modulo the validity of the assumption. Masreliez's assumption is that the state predictor density is Gaussian

$$f(\underline{x}_k|Y^{k-1}) = N(\underline{x}_k; \hat{\underline{x}}_k^{k-1}, M_k) \tag{2.4}$$

where $N(\cdot;\underline{\mu}, C)$ denotes the multivariate normal density with mean vector $\underline{\mu}$ and covariance matrix C. The covariance matrix M_k in (2.4) is the conditional error covariance for predicting x_k

$$M_k = E\{(\underline{x}_k - \hat{\underline{x}}_k^{k-1})(\underline{x}_k - \hat{\underline{x}}_k^{k-1})^T|Y^{k-1}\} . \tag{2.5}$$

It is a very special feature of the completely Gaussian situation that $M_k = M_k(Y$[...] is in fact independent of the data Y^k, and similarly for the conditional filtering error covariance

$$P_k = E\{(\underline{x}_k-\hat{\underline{x}}_k)(\underline{x}_k-\hat{\underline{x}}_k)^T|Y^k\} . \tag{2.6}$$

One should not expect M_k and P_k to be independent of the data Y^k in general, and in fact it turns out that M_k and P_k depend upon the data in an intuitively appealing manner in Masreliez's result, which we now present. It is assumed that the observations are generated by (2.1) - (2.2) with the following known in advance: Φ, H, μ, the covariance matrix Q of the $\underline{\varepsilon}_k$ and the density f_v of the v_k. We may take $\mu = 0$ without loss of generality.

<u>THEOREM</u> (Masreliez) If $f(\underline{x}_k;Y^{k-1}) = N(\underline{x}_k;\hat{\underline{x}}_k^{k-1}, M_k)$, $k \geq 1$, then $\hat{\underline{x}}_k = E\{\underline{x}_k|Y^k\}$, $k \geq 1$, is generated by the recursions

$$\hat{\underline{x}}_k = \hat{\underline{x}}_k^{k-1} + M_k H^T \Psi_k(y_k) \tag{2.7}$$

$$M_{k+1} = \Phi P_k \Phi^T + Q \tag{2.8}$$

$$P_k = M_k - M_k H^T \Psi_k'(y_k)\, H\, M_k \tag{2.9}$$

where

$$\Psi_k(y_k) = -(\partial/\partial y_k)\ell og f_y(y_k|Y^{k-1}) \tag{2.10}$$

is the "score" function for the observation prediction density $f_y(y_k|Y^{k-1})$, $\hat{\underline{x}}_k^{k-1} = \Phi\, \hat{\underline{x}}_{k-1}$ and

$$\Psi_k'(y_k) = -(\partial/\partial y_k)\Psi_k(y_k) . \tag{2.11}$$

<u>Comment 1</u> Masreliez did not specify initial conditions for the above recursions. However the appropriate $\hat{\underline{x}}_1$ and M_1 are given by $\hat{\underline{x}}_1 = E\underline{x}_1 = 0$ and $M_1 = E\{\underline{x}_1\underline{x}_1^T\} = C_x$, i.e., the unconditional mean and covariance of $\underline{x}_1$. The latter satisfies the equation $C_x = \Phi\, C_x \Phi^T + Q$ under stationarity.

<u>Comment 2</u> The observation prediction density $f_y(y_k|Y^{k-1})$ is obtained by convolving the prediction density $f_z(z_y|Y^{k-1}) = N(z_k;H\hat{\underline{x}}_k^{k-1}, HM_kH^T)$ for $z_k = H\underline{x}_k$ with the observation noise density f_v. For autoregressive models H = (1, 0, ..., 0), and $f(z_k|Y^{k-1})$ is just the marginal state prediction density $f(x_k|Y^{k-1})$ for the first component x_k of $\underline{x}_k$.

<u>Masreliez's Filter for Autoregressions</u> Although the detailed structure of Ψ_k and Ψ_k' are not exactly obvious in the theorem statement, an easy calculation reveals the details for the case of pth-order autoregressions. Following the second comment above we note that if $f(\underline{x}_k|Y^{k-1}) = N(\underline{x}_k;\hat{\underline{x}}_k^{k-1}, M_k)$ then the marginal density for the first component x_k of $\hat{\underline{x}}_k$ is

$$f_x(x_k|Y^{k-1}) = N(x_k;\bar{x}_k, m_{1k}) \tag{2.12}$$

where

$$\bar{x}_k = (\hat{\underline{x}}_k^{k-1})_1 = (\Phi\, \hat{\underline{x}}_{k-1})_1 = \sum_{j=1}^{p} \phi_j (\hat{\underline{x}}_{k-1})_j \tag{2.13}$$

is the first component of $\hat{\underline{x}}_k^{k-1}$, $(\hat{\underline{x}}_{k-1})_j$ is the j-th component of $\hat{\underline{x}}_{k-1}$, $1 \le j \le p$ and m_{1k} is the 1-1 element of M_k.

Now suppose that the v_k have the contaminated normal distribution

$$CN(v_k;\gamma,\sigma_o^2,\sigma^2) = (1-\gamma)N(v_k;0,\sigma_o^2) + \gamma\, N(v_k;0,\sigma^2)\ . \tag{2.14}$$

Then since $\bar{y}_k \triangleq E\{y_k|Y^{k-1}\} = E\{x_k|Y^{k-1}\} = \bar{x}_k$, convolution of f_x with $f_v = CN(\gamma,\sigma_o^2,\sigma^2)$ gives

$$f(y_k|Y^{k-1}) = (1-\gamma)N(y_k;\bar{y}_k,\sigma_{ok}^2) + \gamma\, N(y_k;\bar{y}_k,\sigma_k^2) \tag{2.15}$$

with $\sigma_{ok}^2 = m_{1k} + \sigma_o^2$ and $\sigma_k^2 = m_{1k} + \sigma^2$. With $\tilde{y}_k = y_k - \bar{y}_k$ and a slight abuse of notation, this gives

$$\psi_k(\tilde{y}_k) = \frac{\tilde{y}_k}{\sigma_{ok}^2}\left[1 - \left(1 - \frac{\sigma_{ok}^2}{\sigma_k^2}\right) b_k(\tilde{y}_k)\right] \tag{2.16}$$

where

$$b_k(t) = \left[1 + \frac{1-\gamma}{\gamma}\frac{\sigma_k}{\sigma_{ok}}\exp\{-(\sigma_{ok}^{-2} - \sigma_k^{-2})\frac{t^2}{2}\}\right]^{-1} .$$

It may be noted that both $\psi_k(\tilde{y}_k)$ and $\psi_k'(\tilde{y}_k)$ are scalar-valued. Thus noting that $H = (1, 0, \ldots, 0)$ and denoting the first column of M_k by $\underline{m}_k$, the pth-order autoregression versions of (2.7) - (2.9) are

$$\hat{\underline{x}}_k = \Phi\, \hat{\underline{x}}_{k-1} + \underline{m}_k\, \psi_k(\tilde{y}_k) \tag{2.17}$$

$$M_{k+1} = \Phi\, P_k\, \Phi^T + Q \tag{2.18}$$

$$P_k = M_k - \psi_k'(\tilde{y}_k)\, \underline{m}_k\, \underline{m}_k^T \tag{2.19}$$

where the p x p covariance matrix

$$Q = \left(\begin{array}{c|c} \sigma_\varepsilon^2 & \underline{0}^T \\ \hline \underline{0} & 0 \end{array}\right) \tag{2.20}$$

for $\underline{\varepsilon}_k$ is singular.

The shapes of $\psi_k(\tilde{y}_k)$ for two combinations of values of γ and the ratio σ_k/σ_{ok} are shown in Figures 1a and 1b. The corresponding shapes of $\psi_k'(\tilde{y}_k)$ are shown in Figures 1c and 1d. The effect of small, intermediate and large observation-predicti residuals $\tilde{y}_k$ on $\hat{\underline{x}}_k$ and P_k is intuitively appealing. If $\tilde{y}_k$ and σ_k/σ_{ok} are appropriat large then $\hat{x}_k \cong \Phi \hat{x}_{k-1}$ and $P_k \cong M_k$. That is, the filtered value at time k is essentially the prediction based on the filtered value at time k-1, and correspondingly the filtering error-covariance matrix is appropriately set very nearly at the value of the one-step ahead prediction-error covariance matrix.

The shapes in Figures 1a and 1b are strikingly similar, and further calculations revealed that the shape of $\Psi_k(y_k)$ is relatively constant for $.01 \le \gamma \le .1$, and a wide range of values for σ_k/σ_{ok} in the following sense: (i) the symmetrically positioned stationary points of $\Psi_k(\tilde{y}_k)$ nearest to the origin have relatively constant locations of approximately $\pm 3\ \sigma_{ok}$; (ii) the heights of the associated local maximum and minimum are relatively constant at $\pm 2.4\ \sigma_{ok}^{-1}$; and (iii) the width of the transition region between the two pairs of stationary points is roughly $1.0\ \sigma_{ok}$ - $2.0\ \sigma_{ok}$. In fact when $\sigma_k >> \sigma_{ok}$, $\Psi_k(\tilde{y}_k)$ behaves rather like a smooth-rejection rule, except of course for the small but positive slope σ_k^{-2} for $|\tilde{y}_k|$ large (see also Hampel's psi-function ψ_{HA} given by 2.26).

In general it is not clear how to define a scale parameter for the distribution obtained by convolving a Gaussian and a non-Gaussian distribution in terms of the scale parameters of the latter distributions. <u>However, an important consequence of (i) and (ii) above is that we can none-the-less express $\Psi_k(\tilde{y}_k)$ in the following approximate form</u>:

$$\Psi_k(\tilde{y}_k) \cong \frac{1}{\sigma_{ok}} \Psi\left(\frac{\tilde{y}_k}{\sigma_{ok}}\right) \tag{2.21}$$

where Ψ is obtained from (2.16) by setting $\sigma_{ok}^2 = 1$ and σ_k^2 at a fixed and appropriately large value.

<u>ACM Type Filters for Autoregressions</u> The preceeding observation suggests to define an approximate conditional-mean (ACM) <u>type</u> filter for autoregressions in the following way. Let ψ be any bounded and continuous function chosen with robustness considerations in mind, and let

$$s_k^2 = m_{1k} + \sigma_o^2 \tag{2.22}$$

where σ_o^2 is the variance of a nominal Gaussian distribution for the observation noise v_k. Then relying on the structure of Ψ_k as indicated by (2.21), the recursions (2.17) and (2.19) are replaced by

$$\hat{x}_k = \Phi\hat{x}_{k-1} + \frac{m_k}{s_k^2} s_k \psi\left(\frac{y_k}{s_k}\right) \tag{2.23}$$

and

$$P_k = M_k - \psi'\left(\frac{\tilde{y}_k}{s_k}\right) \frac{m_k m_k^T}{s_k^2} . \tag{2.24}$$

The recursion (2.18) and the definition of Q remain the same.

The terminology ACM-filter parallels that of P. Hubers (1964, 1973) introduction of the term M-estimate to denote a maximum-likelihood type estimate for location or regression.

The heuristic motivations for specifying that ψ be bounded and continuous are similar to the reasons for using bounded and continuous psi-functions for M-estimates. Boundedness insures that y_k can not have an unbounded influence on $\hat{\underline{x}}_j$, $k \leq j \leq n$. Continuity insures that small changes (e.g., due to rounding) in y_k do not produce large changes in $\hat{\underline{x}}_j$, $k \leq j \leq n$.

For location parameter problems there is a basic theoretical motivation for requiring that the ψ used in an M-estimate be bounded and continuous. Namely it insures that the estimate is qualitatively robust in Hampel's (1971) sense. Unfortunately we do not yet have a theory of qualitative robustness for filters and smoothers, and thus we rely for now on the above heuristic argument.

Two psi-functions (among others) which are likely candidates for use in ACM-filters are Huber's monotone function

$$\psi_H(t) = \begin{cases} t, & |t| \leq 1 \\ \mathrm{SGN}(t), & |t| > 1 \end{cases} \tag{2.25}$$

and Hampel's two-part redescending function

$$\psi_{HA}(t) = \begin{cases} t & |t| \leq a \\ \frac{a}{b-a}(b-t) & a < t \leq b \\ -\frac{a}{b-a}(b+t) & -b \leq t < -a \\ 0 & |t| > b \;. \end{cases} \tag{2.26}$$

There is a peculiarity about the ACM-filter based on ψ_H in that $|\tilde{y}_k| > s_k$ results in $\psi_H'(\tilde{y}_k/s_k) = 0$ so that $P_k = M_k$, while at the same time $\hat{\underline{x}}_k$ is not equal to the prediction $\Phi\hat{\underline{x}}_{k-1}$ as one might expect if $P_k = M_k$.

On the other hand the ACM-filter based on ψ_{HA} has the intuitively appealing feature that $\hat{\underline{x}}_k = \Phi\hat{\underline{x}}_{k-1}$ and $P_k = M_k$ by virture of $\psi_{HA}(\tilde{y}_k/s_k) = \psi_{HA}'(\tilde{y}_k/s_k) = 0$ when $|\tilde{y}_k| > b \cdot s_k$. This would appear to be the natural embodiment of an outlier-rejection rule in the filtering context. There is however an unusual feature associated with the use of ψ_{HA}. Since $\psi_{HA}'(\tilde{y}_k/s_k) = -a/(b-a)$ when $a \cdot s_k < |\tilde{y}_k| < b \cdot s_k$, the filtering-error variance $(P_k)_{11}$ for x_k is larger than the prediction-error variance m_{1k} in such a case. This contrasts sharply with the fact that $(P_k)_{11}$ is less than m_{1k} when $|\tilde{y}_k| < a \cdot s_k$ and $P_k = M_k$ for $|\tilde{y}_k| > b \cdot s_k$. A possible explanation for such behavior is that when a and b have values of about 2.5 and 3.5 respectively, the intervals $(-bs_k, -as_k)$ and (as_k, bs_k) are those which present a considerable difficulty for determining whether or not $\tilde{y}_k$ contains an outlier. We shall use ψ_{HA} in an example later on.

<u>One-Sided Outlier Interpolation</u> An important use of ACM-filters is for time series situations where the nominal Gaussian distribution of v_k has variance $\sigma_o^2 = 0$. Thus using (2.14) as a point of departure, the time series is assumed to be observed perfectly a large fraction $1-\gamma$ of the time, and observed with an additive error a fraction γ of the time. This "error" need not be an observational error in the usual sense. It may be simply an effect due to any one of a variety of causes which alters the autoregressive structure locally in time in an (approximately) additive manner. With $\sigma_o^2 = 0$, $s_k^2 = m_{1k}$, and recursions (2.23), (2.24) and (2.18) become

$$\underline{\hat{x}}_k = \Phi\underline{\hat{x}}_{k-1} + \underline{\tilde{m}}_k s_k \psi(\tilde{y}_k/s_k) \tag{2.27}$$

$$P_k = M_k - \psi'(\tilde{y}_k/s_k) \underline{\tilde{m}}_k \underline{\tilde{m}}_k^T \tag{2.28}$$

$$M_{k+1} = \Phi P_k \Phi^T + Q \tag{2.29}$$

where

$$\underline{\tilde{m}}_k^T = (1, \frac{m_{2k}}{m_{1k}}, \frac{m_{3k}}{m_{1k}}, \ldots, \frac{m_{pk}}{m_{1k}}) . \tag{2.30}$$

Now $\hat{x}_k = (\underline{\hat{x}}_k)_1 = (E\{\underline{x}_k|Y^k\})_1 = E\{x_k|Y^k\}$ is the filtered value at time k based on the vector estimate $\underline{\hat{x}}_k$, and $\bar{y}_k = \sum_{j=1}^p \phi_j (\underline{x}_{k-1})_j$ is the prediction of y_k based on Y^{k-1}. Suppose $\psi = \psi_{HA}$. If the prediction residual $\tilde{y}_k = y_k - \bar{y}_k$ satisfies $|\tilde{y}_k| > b$ then $\hat{x}_k$ is just the predicted value $\hat{x}_k = \bar{y}_k$, and if $|\tilde{y}_k| < a$ then $\hat{x}_k = \bar{y}_k + (y_k - \bar{y}_k) = y_k$. When a time series contains only a rather small fraction of outliers, and the constants a and b are appropriately adjusted, we find that $\hat{x}_k = y_k$ a large fraction of the time and $\hat{x}_k \cong \bar{y}_k$ a small fraction of time. It seems appropriate to refer to an ACM type filter operating under such conditions as a <u>one-sided interpolator</u>.

<u>A Question of Simplification</u> Thomson (1977) and Kleiner, Martin, and Thomson (1979) made use of the computationally simpler version of filter

$$\hat{x}_k = \sum_{j=1}^p \phi_j \hat{x}_{k-j} + s\psi(\frac{y_k - \sum_{j=1}^p \phi_j \hat{x}_{k-j}}{s}) \tag{2.31}$$

where s and the ϕ_j are estimated from the data. This filter, or one-side outlier interpolator, differs from the ACM version (2.27) - (2.28) in two respects. First of all s is a globablly determined estimate which does not depend upon the data in the intuitively appealing local manner that s_k does. Secondly, even if s in (2.31) were replaced by the s_k generated by (2.27) - (2.29), the filter sequence produced by (2.31) would not be the same as that produced by (2.27) - (2.29). For the latter ACM-filter makes use of the covariance structure of $\underline{\tilde{m}}_k$ to correct the entire prediction vector $\Phi\underline{\hat{x}}_{k-1}$ at each step of the recursion.

Although the second difference would seem to favor the ACM-filter over the simple filter (2.31), it is not yet clear to me whether the ensuing filter performances vary greatly due to this difference. The first difference mentioned above appears to be more crucial. For the simple filter can lose track of the data, never to regain it, particularly if ψ redescends. On the other hand the ACM-filter has the property that s_k increases during time periods for which the filter has lost track, thereby eventually allowing the filter to regain track of the data. An illustrative example is given in Martin (1979b).

<u>The Gaussian Assumption</u> One might well ask why it is worthwhile to propose the class of ACM-filters when the Gaussian assumption on the state prediction density $f(\underline{x}_k|Y^{k-1})$ appears questionable. In fact it is shown in Section 5 that if both $f(\underline{x}_k|Y^{k-1})$ and the unconditional distribution of $\underline{x}_k$ are assumed to be Gaussian then the distribution of v_k must be Gaussian.

On the other hand Monte Carlo simulations for first order autoregressions by Masreliez (1975) and Martin and DeBow (1976) indicate that ACM-filters are very close to being optimal. Furthermore heuristic arguments indicate that $f(\underline{x}_k|Y^{k-1})$ should be approximately Gaussian. In Section 5 a continuity property of $f(\underline{x}_k|Y^{k-1})$ is established which lends further credence to the belief that this density is nearly Gaussian when v_k has a heavy-tailed non-Gaussian distribution.

3. APPROXIMATE CONDITIONAL-MEAN SMOOTHING AND OUTLIER INTERPOLATION

It now turns out to be rather easy to construct ACM type smoothers using the ACM type filters described in the previous section. The basis of doing so is a theorem which yields the structure of an ACM-smoother in terms of an ACM-filter. This result is a companion of Masreliez's theorem in that it also makes use of the assumption that the state-prediction density $f(\underline{x}_k|Y^{k-1})$ is Gaussian. A conditional-mean smoothed value is denoted $\hat{\underline{x}}_k^n = E\{\underline{x}_k|Y^n\}$, $1 \le k < n$, where the conditioning is now on all the data. For $k = n$ we have $\hat{\underline{x}}_n^n = \hat{\underline{x}}_n$ which is a filtered value.

<u>THEOREM (ACM-SMOOTHER)</u> Suppose that $f(\underline{x}_k|Y^{k-1}) = N(\underline{x}_k;\hat{\underline{x}}_k^{k-1}, M_k)$ where $\hat{\underline{x}}_k^{k-1} = \Phi\hat{\underline{x}}_{k-1}$ and $\hat{\underline{x}}_k = E\{\underline{x}_k|Y^k\}$, $1 \le k \le n$, is the ACM-filter of Masreliez's theorem, with the appropriate initial conditions. Then assuming M_{k+1}^{-1} exists, $\hat{\underline{x}}_k^n$ satisfies the backward recursion

$$\hat{\underline{x}}_k^n = \hat{\underline{x}}_k + P_k\Phi^T M_{k+1}^{-1} \cdot (\hat{\underline{x}}_{k+1}^n - \hat{\underline{x}}_{k+1}^k), \quad 1 \le k \le n-1 , \tag{3.1}$$

with initial condition $\hat{\underline{x}}_n^n = \hat{\underline{x}}_n$. The smoothing-error covariance matrix

$$P_k^n = E\{(\hat{\underline{x}}_k - \hat{\underline{x}}_k^n)(\hat{\underline{x}}_k - \hat{\underline{x}}_k^n)^T|Y^n\}$$

satisfies the backward recursion

$$P_k^n = P_k + A_k(P_{k+1}^n - M_{k+1})A_k^T \tag{3.2}$$

with initial condition $P_n^n = P_n$ and

$$A_k = P_k\Phi^T M_{k+1}^{-1} . \tag{3.3}$$

A proof of the above theorem is deferred to Section 4.

Comment 1 The form of the ACM-smoothing recursions are identical to those obtained by Meditch (1967, 1969) using either a linearity assumption or a Gaussian assumption. This fact, for which I had held out a little hope, is rather pleasing. The smoothed values are obtained by first computing $\hat{\underline{x}}_1, \ldots, \hat{\underline{x}}_n$ from (2.7) - (2.9), and then computing $\hat{\underline{x}}_{n-1}^n, \hat{\underline{x}}_{n-2}^n, \ldots, \hat{\underline{x}}_1^n$, in that order from (3.1).

Comment 2 Expressing (3.1) in the form

$$\hat{\underline{x}}_k^n = (I-A_k\Phi)\,\hat{\underline{x}}_k + (A_k\Phi)\,\Phi^{-1}\hat{\underline{x}}_{k+1}^n \qquad (3.4)$$

yields a more intuitive view of $\hat{\underline{x}}_k^n$ as a weighted combination of the ACM-filtered value at time k and the one-step backwards prediction $\Phi^{-1}\hat{\underline{x}}_{k+1}^n$ of $\underline{x}_k$ based on the ACM-smoothed value at time k+1. The weight matrices sum to I and it is easy to check that $0 \le A_k\Phi \le I$.

ACM-Type Smoothers An approximate conditional-mean type smoother is obtained from (3.1) by using an ACM type filter as discussed in Section 2, to obtain the $\hat{\underline{x}}_k$, $1 \le k \le n$.

Filter and Smoother Output Choices When $\underline{x}_k$ is an autoregression of order $p \ge 2$ there is a question concerning which coordinate of $\hat{\underline{x}}_k$ or $\hat{\underline{x}}_k^n$ should be used as the filter or smoother output. This question was implicitly answered for the filter in the context of the one-sided interpolator discussion of the preceeding section. For the general ACM-filter context the same choice, namely the first coordinate $\hat{x}_k = (\hat{\underline{x}}_k)_1$ of $\hat{\underline{x}}_k$ gives the conditional-mean $E\{x_k|Y^k\}$ for filtering. However we could also make the choice $\hat{x}_k = (\hat{\underline{x}}_{k+p-1})_p$, the last coordinate of $\hat{\underline{x}}_k$. Since $(\hat{\underline{x}}_{k+p-1})_p = E\{x_k|Y^{k+p-1}\}$ this results in a "fixed lag" smoother of lag p-1.

For ACM-smoothers the first coordinate $\hat{x}_k^n = (\hat{\underline{x}}_k^n)_1 = (E\{\underline{x}_k|Y^n\})_1$ yields the desired conditional-mean $E\{x_k|Y^n\}$. It may be noted that when (3.1) is used for $k = n-1, n-2, \ldots, 1$ in succession, the final vector smooth is $(\hat{\underline{x}}_1^n)^T = (\hat{x}_1^n, \hat{x}_0^n, \ldots, \hat{x}_{-p+2})$ which yields $\hat{x}_1^n$. Thus the smoother goes to both ends of the data in a natural way.

Outlier Interpolation and σ_0^2 as a Smoothness Parameter The ACM-smoother becomes a (two-sided) outlier interpolator when σ_0^2 is set at zero and the ACM-filter is a one-sided outlier interpolator as discussed earlier.

It appears that σ_0^2 may also be used as a smoothness parameter to control the degree of smoothness of the output. In the Gaussian case σ_0^2 certainly has this interpretation. $\sigma_0^2 = 0$ corresponds to no observation noise and no smoothing, whereas large values of σ_0^2 correspond to large observation noise and correspondingly high degree of smoothing. The behavior of ACM-filters and smoothers based on a redescending psi-function, such as ψ_{HA}, as a function of σ_0^2 is similar. When $\sigma_0^2 = 0$ in such circumstances the phrase "no smoothing" is changed to "little or no alteration of most of the data points."

However there is an apparent difficulty in using σ_0^2 as a smoothing parameter in that smoothness and robustness seem to be incompatible goals. For increasing σ_0^2 to achieve more smoothness results in increasing the scale parameter $s_k = (m_{1k}+\sigma_0^2)^{1/2}$

used in the psi-function which operates on the prediction residuals. This would seem to give less protection against outliers. A potential way around the difficulty is suggested next.

The Linear Part of ACM-Smoothers It was mentioned in the introduction that Mallows (1979a) obtained a useful decomposition of non-linear smoothers into the sum of a "linear part" and a residuals process which are mutually orthogonal. Mallows also gave the following result (Theorem 4.6 of the cited reference): if a non-linear smoother is followed by a linear smoother then the linear part of the concatenation is the same as the linear part of the nonlinear smoother followed by the linear smoother. He then comments that "This property holds out hope of greatly simplifying the task of designing a robust smoother; one just uses a nonlinear smoother to achieve the desired insensitivity to outliers, and follows it with a linear filter to achieve a desired transfer shape."

I believe that well designed outlier-interpolator forms of ACM-smoothers will be quite helpful in achieving a more specific version of the above separation principle. Namely, they will achieve insensitivity to outliers without changing the spectrum corresponding to the "good" data very much at all. In other words the "linear part" (assuming Mallows' concept applies in the same way to ACM-smoothers)[†] will be very nearly the identity operator. Then the output of the outlier-interpolator can be followed by a linear smoother whose transfer function in fact has very nearly the desired overall shape.

It is important to note that a first-stage processor having a linear part which is very nearly the identity operator is also highly desirable if one wishes to estimate the parameters of autoregressive-moving-average models in a robust manner using conventional methodology at the second stage.

Parameter Estimation and Implementation of ACM-Smoothers In order to implement ACM-smoothers for autoregressions we need robust estimates of the parameters μ, σ_ε, $\phi^T = (\phi_1,\dots,\phi_p)$ and the first column of the p x p covariance matrix C_x of $\underline{x}_k$. The latter is needed for initializing the ACM-filter recursion.

For simplicity μ may be estimated robustly with an ordinary location M-estimate $\hat{\mu}$. Translation equivariance for the smoother is then easily obtained by using $y_i - \hat{\mu}$ in place of the y_i, then obtaining the smoothed values $\hat{x}_k^n$, and finally replacing the $\hat{x}_k^n$ with $\hat{x}_k^n + \hat{\mu}$.

One way to estimate ϕ and σ_ε is by using a generalized M-estimate (GM-estimate) for autoregressions as described by Denby and Martin (1979), Martin (1979c), Martin and Zeh (1979). The GM-estimate yields a robust estimate of C_x^{-1} in a factored form which allows for convenient inversion to obtain an estimate of the first column of C_x.

† Unfortunately Mallows' decomposition does not apply to ACM-smoothers since one of the assumptions of his theory is that the smoother have a fixed span. (A smoother $\hat{x}_k$ has fixed span $S = 2L+1$ if $\hat{x}_k$ depends upon y_{k+j}, $-L \le j \le L$, except possibly for end effects.) Recursive filters and smoothers of the ACM variety do not have fixed spans, and I do not yet know whether a similar decomposition can be established for them.

Another possibility is to use an iterative procedure starting with the usual least-squares estimates $\hat{\phi}$ and $\hat{\sigma}_\varepsilon^2$, the first column of $\hat{C}_x$ being obtained from these estimates. Then the $\hat{x}_k^n$ are obtained using $\hat{\phi}$, $\hat{\sigma}_\varepsilon^2$ and $\hat{C}_x$ in place of ϕ, σ_ε^2 and C_x, and these $\hat{x}_k^n$ are used to obtain new least-squares estimates. This procedure is iterated until relative convergence hopefully sets in.

Little is known about the convergence properties of this method. However the successful use of such a procedure in which the simple filter (2.31) was used instead of an ACM-smoother was reported by Kleiner, Martin, and Thomson (1979). A similar procedure using an ACM-filter instead of simple robust filter produces an approximate-conditional-mean M-estimate of $\phi_1, \ldots, \phi_p$ as described by Martin (1979a). This estimate may be interpreted as an approximate non-Gaussian maximum-likelihood type estimate of the parameters (Martin, 1979c). Although it does not seem likely that such an interpretation is possible when ACM-smoothed values are used instead of filtered values, the procedure is still appealing and its efficacy should be investigated.

An Example Figure 2a displays the "suspended bank deposits" time series used as an example by J. W. Tukey (1977a) to illustrate the use of a particular L-smoother. This data was smoothed using a first-order autoregression ACM-smoother based on the Hampel two-part redescending psi-function ψ_{HA} given by (2.26) with a = 2.6 and b = 3.6. The original data y_i was replaced by the centered data $y_i - \hat{\mu}$ with $\hat{\mu}$ an ordinary location M-estimate using Tukey's (1977b) bisquare psi-function. The robust estimates $\hat{\phi} = .97$ and $\hat{\sigma}_\varepsilon = 45.8$ were obtained using a first order GM-estimate as described in Denby and Martin (1979). A robust scale estimate $\hat{\sigma}_x$ was used to obtain the initial condition for the forward filtering portion of the smoother by setting $M_1 = \hat{\sigma}_x^2$.

The value $\sigma_0^2 = 800$ was used, which yields a small amount of smoothing in addition to outlier interpolation. The result, shown in Figure 2b, appears to be a reasonable smooth of the suspended bank deposits data (cf., Tukey, 1977a, pg. 246).

The estimated value $\hat{\phi} = .97$ is rather close to the value $\phi = 1$ for a random walk, and this seems reasonable in view of the nonstationary appearance of the time series. The large jump might well be roughly attributed to an ε_k outlier. The other outliers may be attributable to nonzero values of v_k, with the majority of the v_k having essentially zero values.

This example suggests that a useful and simple robust smoother might be obtained for nonstationary appearing time series by simply setting $\phi = 1$. Then σ_ε could be estimated by robustly estimating the scale of the first differences $\Delta y_k = y_k - y_{k-1}$. A robust estimate of σ_x will be required as an initial condition.

4. PROOF OF THE ACM-SMOOTHING THEOREM

In order to prove the ACM-Smoothing Theorem, we shall make use of the following lemma concerning the derivative of $\ell og f(Y_{k+1}^n | Y^k)$ with respect to the last conditioning variable y_k. All necessary regularity conditions are assumed for the proofs.

Lemma 1 If $f(\underline{x}_k|Y^{k-1}) = N(\underline{x}_k-\underline{x}_k^{k-1};0,M_k)$ then

$$(\partial/\partial y_k)\log f(Y_{k+1}^n|Y^k) = G_k HM_k \Phi^T M_{k+1}^{-1} \cdot (\hat{\underline{x}}_{k+1}^n - \hat{\underline{x}}_{k+1}^k) .$$

Proof

$$f(Y_{k+1}^n|Y^k) = \int f(\underline{x}_{k+1}|Y^k) f(Y_{k+1}^n|\underline{x}_{k+1},Y^k) d\underline{x}_{k+1}$$

$$= \int f(\underline{x}_{k+1}|Y^k) f(Y_{k+1}^n|\underline{x}_{k+1}) d\underline{x}_{k+1}$$

and so

$$(\partial/\partial y_k) f(Y_{k+1}^n|Y^k) = \int (\partial/\partial y_k) f(\underline{x}_{k+1}|Y^k) \cdot f(Y_{k+1}^n|\underline{x}_{k+1}) d\underline{x}_{k+1} .$$

But

$$(\partial/\partial y_k) f(\underline{x}_{k+1}|Y^k) = (\partial/\partial y_k) N(\underline{x}_{k+1}-\hat{\underline{x}}_{k+1}^k;0,M_k)$$

$$= (\partial/\partial y_k)\hat{\underline{x}}_k^k \cdot (\partial/\partial \hat{\underline{x}}_k^k) N(\underline{x}_{k+1}-\hat{\underline{x}}_{k+1}^k;0,M_k)$$

and

$$(\partial/\partial y_k)\hat{\underline{x}}_k^k = (\partial/\partial y_k)\{\underline{x}_k^{k-1}+M_k H^T g_k(y_k)\} = (\partial/\partial y_k) g_k(y_k) \cdot HM_k = G_k HM_k .$$

Thus

$$(\partial/\partial y_k) f(\underline{x}_{k+1}|Y^k) = -G_k HM_k \Phi^T (\partial/\partial \underline{x}_{k+1}) f(\underline{x}_{k+1}|Y^k)$$

where we have used the fact that

$$(\partial/\partial \hat{\underline{x}}_k^k)\hat{\underline{x}}_{k+1}^k = (\partial/\partial \underline{x}_k^k)\Phi \hat{\underline{x}}_k^k = \Phi^T .$$

This gives

$$(\partial/\partial y_k) f(Y_{k+1}^n|Y^k) = -G_k HM_k \Phi^T \int (\partial/\partial \underline{x}_{k+1}) f(\underline{x}_{k+1}|Y^k) \cdot f(Y_{k+1}^n|\underline{x}_{k+1}) d\underline{x}_{k+1}$$

$$= G_k HM_k \Phi^T M_{k+1}^{-1} \int (\underline{x}_{k+1}-\hat{\underline{x}}_{k+1}^k) f(\underline{x}_{k+1}|Y^k) \cdot f(Y_{k+1}^n|\underline{x}_{k+1},Y^k) d\underline{x}_{k+1}$$

Thus

$$(\partial/\partial y_k)\log f(Y_{k+1}^n|Y^k) = \{f(Y^k)/f(Y^n)\}(\partial/\partial y_k) f(Y_{k+1}^n|Y^k)$$

$$= G_k HM_k \Phi^T M_{k+1}^{-1} \int (\underline{x}_{k+1}-\hat{\underline{x}}_{k+1}^k) f(\underline{x}_{k+1}|Y^n) d\underline{x}_{k+1}$$

$$= G_k HM_k \Phi^T M_{k+1}^{-1} \cdot (\hat{\underline{x}}_{k+1}^n - \hat{\underline{x}}_{k+1}^k) .$$

Proof of ACM-Smoothing Theorem

We first write $\hat{\underline{x}}_k^n = E\{\underline{x}_k|Y^n\}$ in the form

$$\hat{\underline{x}}_k = \int \underline{x}_k f(\underline{x}_k|Y^n)d\underline{x}_k = \hat{\underline{x}}_k^{k-1} + \iint(\underline{x}_k-\hat{\underline{x}}_k^{k-1})f(\underline{x}_k|Y^n)d\underline{x}_k .$$

Noting that

$$\begin{aligned} f(\underline{x}_k|Y^n) &= f(\underline{x}_k,Y_{k+1}^n|Y^k)f(Y^k)/f(Y^n) \\ &= f(Y_{k+1}^n|Y^k,\underline{x}_k)f(\underline{x}_k|Y^k)f(Y^k)/f(Y^n) \\ &= f(Y_{k+1}^n|\underline{x}_k)f(\underline{x}_k|Y^k)f(Y^k)/f(Y^n) \end{aligned}$$

we have

$$\hat{\underline{x}}_k = \hat{\underline{x}}_k^{k-1} + \int(\underline{x}_k-\hat{\underline{x}}_k^{k-1})f(Y_{k+1}^n|\underline{x}_k)f(\underline{x}_k|Y^k)\{(f(Y^k)/f(Y^n)\}d\underline{x}_k$$

Since $f(\underline{x}_k|Y^{k-1}) = N(\underline{x}_k;\hat{\underline{x}}_k^{k-1},M_k)$ by assumption, Masreliez's integration-by-parts device based on the identity

$$(\underline{x}_k-\hat{\underline{x}}_k^{k-1})f(\underline{x}_k|Y^{k-1}) = -M_k(\partial/\partial\underline{x}_k)f(\underline{x}_k|Y^{k-1})$$

may be used to obtain

$$\begin{aligned} \hat{\underline{x}}_k^n &= \hat{\underline{x}}_k^{k-1} + \{f(Y^k)/f(Y^n)\}\int f(Y_{k+1}^n|\underline{x}_k)(\underline{x}_k-\hat{\underline{x}}_k^{k-1})f(\underline{x}_k|Y^{k-1})\{f_v(y_k-H\underline{x}_k)/f(y_k|Y^{k-1})\}d\underline{x}_k \\ &= \hat{\underline{x}}_k^{k-1} - \{f(Y^k)/f(Y^n)\}M_k\int f(Y_{k+1}^n|\underline{x}_k)\{f_v(y_k-H\underline{x}_k)/f(y_k|Y^{k-1})\}(\partial/\partial\underline{x}_k)f(\underline{x}_k|Y^{k-1})d\underline{x}_k \\ &= \hat{\underline{x}}_k^{k-1} + \{f(Y^k)/f(Y^n)\}M_k\int(\partial/\partial\underline{x}_k)\{f(Y_{k+1}^n|\underline{x}_k)f_v(y_k-H\underline{x}_k)\}\cdot\{f(\underline{x}_k|Y^{k-1})/f(y_k|Y^{k-1})\}d\underline{x}_k \\ &= \hat{\underline{x}}_k^{k-1} + \{f(Y^k)/f(Y^n)\}M_k\int(\partial/\partial\underline{x}_k)f(Y_{k+1}^n|\underline{x}_k)\cdot f(\underline{x}_k|Y^k)d\underline{x}_k \\ &\quad - \{f(Y^k)/f(Y^n)\}M_kH^T\int f(Y_{k+1}^n|\underline{x}_k)(\partial/\partial y_k)f_v(y_k-H\underline{x}_k)\cdot\{f(\underline{x}_k|Y^{k-1})/f(y_k|Y^{k-1})\}d\underline{x}_k \\ &= \hat{\underline{x}}_k^{k-1} + T_1 + T_2 . \end{aligned}$$

Now

$$\begin{aligned} (\partial/\partial y_k)f(\underline{x}_k|Y^k) &= (\partial/\partial y_k)f_v(y_k-H\underline{x}_k)f(\underline{x}_k|Y^{k-1})/f(y_k|Y^{k-1}) \\ &= (\partial/\partial y_k)f_v(y_k-H\underline{x}_k)\cdot f(\underline{x}_k|Y^{k-1})/f(y_k|Y^{k-1}) \\ &\quad - f_v(y_k-H\underline{x}_k)\{f(\underline{x}_k|Y^{k-1})/f^2(y_k|Y^{k-1})\}(\partial/\partial y_k)f(y_k|Y^{k-1}) \end{aligned}$$

and using this in T_2 gives

$$T_2 = -\{f(Y^k)/f(Y^n)\}M_kHT\int f(Y^n_{k+1}|\underline{x}_k)(\partial/\partial y_k)f(\underline{x}_k|Y^k)d\underline{x}_k$$

$$- \{f(Y^k)/f(Y^n)\}M_kH^T\int f(Y^n_{k+1}|\underline{x}_k)f(\underline{x}_k|Y^k)(\partial/\partial y_k)\{f(y_k|Y^{k-1})\}f(y_k|Y^{k-1})d\underline{x}_k$$

$$= - \{f(Y^k)/f(Y^n)\}M_kH^T\int f(Y_{k+1}|\underline{x}_k)(\partial/\partial y_k)f(\underline{x}_k|Y^k)d\underline{x}_k + M_kH^Tg_k(y_k)$$

$$= -M_kH^T\{f(Y^k)/f(Y^n)\}\int(\partial/\partial y_k)f(x_k,Y^n_{k+1}|Y^k)d\underline{x}_k + M_kH^Tg_k(y_k)$$

$$= -M_kH^T\{f(Y^k)/f(Y^n)\}(\partial/\partial y_k)f(Y^n_{k+1}|Y^k) + M_kH^Tg_k(y_k)$$

$$= -M_kH^T(\partial/\partial y_k)\ell og f(Y^n_{k+1}|Y^k) + M_kH^Tg_k(y_k) .$$

Now use of Lemma 1 gives

$$T_2 = -M_kH^TG_kHM_k\Phi^TM^{-1}_{k+1} \cdot (\hat{\underline{x}}^n_{k+1}-\hat{\underline{x}}^k_{k+1}) + M_kH^Tg_k(y_k) .$$

Thus we have

$$\hat{\underline{x}}^n_k = \hat{\underline{x}}^{k-1}_k + T_1 - M_kH^TG_kHM_k\Phi^TM^{-1}_{k+1} \cdot (\hat{\underline{x}}^n_{k+1}-\hat{\underline{x}}^k_{k+1}) + M_kH^Tg_k(y_k)$$

or

$$\hat{\underline{x}}^n_k = \hat{\underline{x}}^k_k + T_1 - M_kH^TG_kHM_k\Phi^TM^{-1}_{k+1} \cdot (\hat{\underline{x}}^n_{k+1}-\hat{\underline{x}}^k_{k+1})$$

where

$$T_1 = \{f(Y^k)/f(Y^n)\}M\int(\partial/\partial\underline{x}_k)f(Y^n_{k+1}|\underline{x}_k) \cdot f(\underline{x}_k|Y^k)d\underline{x}_k .$$

Since

$$y_{k+j} = H\underline{x}_{k+j} + v_{k+j} = H\Phi^j\underline{x}_k + H\sum_{\ell=0}^{j-1}\Phi^\ell\varepsilon_{k-\ell} + v_{k+j} = H\Phi^j\underline{x}_k + \xi_{k+j}$$

we have

$$f(Y^n_{k+1}|\underline{x}_k) = f(\xi_{k+1},\ldots,\xi_n|\underline{x}_k) = f(y_{k+1}-H\Phi\underline{x}_k,\ldots,y_n-H\Phi^{n-k}\underline{x}_k)$$

and so using the chain rule for functions of several variables gives

$$(\partial/\partial\underline{x}_k)f(Y^n_{k+1}|\underline{x}_k) = (\partial/\partial\underline{x}_k)f(y_{k+1}-H\Phi\underline{x}_k,\ldots,y_n-H\Phi^{n-k}\underline{x}_k)$$

$$= -\sum_{j=1}^{n-k}(\Phi^j)^TH^T(\partial/\partial y_{k+j})f(y_{k+1}-H\Phi\underline{x}_k,\ldots,y_n-H\Phi^{n-k}\underline{x}_k)$$

$$= -\sum_{j=1}^{n-k}(\Phi^j)^TH^T(\partial/\partial y_{k+j})f(y_{k+1},\ldots,y_n|\underline{x}_k) .$$

This allows us to express T_1 as follows:

$$T_1 = M_k\{f(Y^k)/f(Y^n)\}\int(\partial/\partial\underline{x}_k f(y_{k+1},\ldots,y_n|\underline{x}_k)\cdot f(\underline{x}_k|Y^k)d\underline{x}_k$$

$$= -M_k(\phi^j)^T H^T\{f(Y^k)/f(Y^n)\}\int\sum_{j=1}^{n-k}(\partial/\partial y_{k+j})f(y_{k+1},\ldots,y_n|\underline{x}_k)\cdot f(\underline{x}_k|Y^k)d\underline{x}_k$$

$$= -\sum_{j=1}^{n-k} M_k(\phi^j)^T H^T\{f(Y^k)/f(Y^n)\}(\partial/\partial y_{k+j})\int f(y_{k+1},\ldots,y_n|\underline{x}_k)f(\underline{x}_k|Y^k)d\underline{x}_k$$

$$= -\sum_{j=1}^{n-k} M_k(\phi^j)^T H^T\{f(Y^k)/f(Y^n)\}(\partial/\partial y_{k+j})\int f(y_{k+1},\ldots,y_n,\underline{x}_k|Y^k)d\underline{x}_k$$

$$= -\sum_{j=1}^{n-k} M_k(\phi^j)^T H^T(\partial/\partial y_{k+j})\log f(Y^n)\ .$$

T_1 can be expressed in a more useful form by noting that for $1 \le k \le n-1$

$$(\partial/\partial y_{k+j})f(Y^n) = (\partial/\partial y_{k+j})\{f(y_{k+j}|Y^{k+j-1})f(Y^n_{k+j+1}|Y^{k+j})f(Y^{k+j-1})\}$$

$$= (\partial/\partial y_{k+j})f(y_{k+j}|Y^{k+j-1})\cdot f(Y^n_{k+j+1}|Y^{k+j})f(Y^{k+j-1})$$

$$+ f(y_{k+j}|Y^{k+j-1})(\partial/\partial y_{k+j})f(Y^n_{k+j+1}|Y^{k+j})\cdot f(Y^{k+j-1})$$

and so for $1 \le k < n-1$

$$(\partial/\partial y_{k+j})\log f(Y^n) = (\partial/\partial y_{k+j})\log f(y_{k+j}|Y^{k+j-1}) + (\partial/\partial y_{k+j})\log f(Y^n_{k+j+1}|Y^{k+j})$$

$$= -g_{k+j}(y_{k+j}) + (\partial/\partial y_{k+j})\log f(Y^n_{k+j+1}|Y^{k+j})\ .$$

$$= -g_{k+j}(y_{k+j}) + G_{k+j}HM_{k+j}\phi^T M^{-1}_{k+j+1}(\hat{\underline{x}}^n_{k+j+1}-\hat{\underline{x}}^{k+j}_{k+j+1})\ .$$

For $k = n-1$

$$(\partial/\partial y_n)\log f(Y^n) = -g_n(y_n)\ .$$

Thus

$$T_1 = \sum_{j=1}^{n-k} M_k(\phi^j)^T H^T g_{k+j}(y_{k+j}) - \sum_{j=1}^{n-k-1} M_k(\phi^j)^T H^T G_{k+j}HM_{k+j}\phi^T M^{-1}_{k+j+1}\cdot(\hat{x}^n_{k+j+1}-\hat{\underline{x}}^{k+j}_{k+j+1})$$

and so we have

$$\hat{\underline{x}}^n_k = \hat{\underline{x}}^k_k - M_k H^T G_k HM_k\phi^T M^{-1}_{k+1}\cdot(\hat{\underline{x}}^n_{k+1}-\hat{\underline{x}}^k_{k+1}) + \sum_{j=1}^{n-k} M_k(\phi^j)H^T g_{k+j}(y_{k+j})$$

$$- \sum_{j=1}^{n-k-1} M_k(\phi^j)^T H^T G_{k+j}HM_{k+j}\phi^T M^{-1}_{k+j+1}\cdot(\hat{\underline{x}}^n_{k+j+1}-\hat{\underline{x}}^{k+j}_{k+j+1})\ .$$

For $k = n-1$ the last term vanishes leaving

$$\hat{\underline{x}}_{n-1}^{n} = \hat{\underline{x}}_{n-1}^{n-1} + M_{n-1}\{\Phi^T H^T g_n(y_n) - H^T G_{n-1} H M_{n-1}\Phi^T M_n^{-1}(\hat{x}_n^n - \hat{x}_n^{n-1})\}$$

$$= \hat{\underline{x}}_{n-1}^{n-1} + \{M_{n-1} - M_{n-1}H^T G_{n-1} H M_{n-1}\}\Phi^T M_n^{-1} \cdot (\hat{\underline{x}}_n^n - \hat{\underline{x}}_n^{n-1})$$

$$= \hat{\underline{x}}_{n-1}^{n-1} + P_{n-1}\Phi^T M_n^{-1} \cdot (\hat{\underline{x}}_n^n - \hat{\underline{x}}_n^{n-1}) \ .$$

Now the result for $\hat{x}_k^n$ is obtained by induction. Suppose that

$$\hat{\underline{x}}_{k+j}^{n} = \hat{\underline{x}}_{k+j}^{k+j} + P_{k+j}\Phi^T M_{k+j+1}^{-1} \cdot (\hat{\underline{x}}_{k+j+1}^{n} - \hat{\underline{x}}_{k+j+1}^{k+j})$$

for $1 \le k < n-1$, $1 \le j \le n-k-1$. Then for such k we have

$$\hat{\underline{x}}_k^n = \hat{\underline{x}}_k^k + P_k\Phi^T M_{k+1}^{-1} \cdot (\hat{\underline{x}}_{k+1}^n - \hat{\underline{x}}_{k+1}^k) - M_k\Phi^T M_{k+1}^{-1} \cdot (\hat{\underline{x}}_{k+1}^n - \hat{\underline{x}}_{k+1}^k)$$

$$+ M_k\Phi^T M_{k+1}^{-1} \cdot (\hat{\underline{x}}_{k+1}^{k+1} - \hat{\underline{x}}_{k+1}^k) + \sum_{j=1}^{n-k-1} M_k(\Phi^j)^T\Phi^T M_{k+j+1}^{-1} \cdot (\hat{\underline{x}}_{k+j+1}^{k+j+1} - \hat{\underline{x}}_{k+j+1}^{k+j})$$

$$- \sum_{j=1}^{n-k-1} M_k(\Phi^j)^T H^T G_{k+j} H M_{k+j}\Phi^T M_{k+j+1}^{-1} \cdot (\hat{\underline{x}}_{k+j+1}^n - \hat{\underline{x}}_{k+j+1}^{k+j}) \ .$$

But

$$H^T G_{k+j} H M_{k+j}\Phi^T M_{k+j+1}^{-1} \cdot (\hat{\underline{x}}_{k+j+1}^n - \hat{\underline{x}}_{k+j+1}^{k+j}) = \Phi^T M_{k+j+1}^{-1} \cdot (\hat{\underline{x}}_{k+j+1}^n - \hat{\underline{x}}_{k+j+1}^{k+j}) - M_{k+j}^{-1} \cdot (\hat{\underline{x}}_{k+j}^n - \hat{\underline{x}}_{k+}^{k+}$$

and so

$$\hat{\underline{x}}_k^n = \hat{\underline{x}}_k^k + P_k\Phi^T M_{k+1}^{-1} \cdot (\hat{\underline{x}}_{k+1}^n - \hat{\underline{x}}_{k+1}^k) - M_k\Phi^T M_{k+1}^{-1} \cdot (\hat{\underline{x}}_{k+1}^n - \hat{\underline{x}}_{k+1}^k)$$

$$+ M_k\Phi^T M_{k+1}^{-1} \cdot (\hat{\underline{x}}_{k+1}^{k+1} - \hat{\underline{x}}_{k+1}^k) + \sum_{j=1}^{n-k-1} M_k(\Phi^j)^T\Phi^T M_{k+j+1}^{-1} \cdot (\hat{\underline{x}}_{k+j+1}^{k+j+1} - \hat{\underline{x}}_{k+j+1}^{k+j})$$

$$- \sum_{j=1}^{n-k-1} M_k(\Phi^j)^T\{\Phi^T M_{k+j+1}^{-1} \cdot (\hat{\underline{x}}_{k+j+1}^n - \hat{\underline{x}}_{k+j+1}^{k+j}) - M_{k+j}^{-1}(\hat{\underline{x}}_{k+j}^n - \hat{\underline{x}}_{k+j}^{k+j})$$

$$= \hat{\underline{x}}_k^k + P_k\Phi^T M_{k+1}^{-1} \cdot (\underline{x}_{k+1}^n - \hat{\underline{x}}_{k+1}^k) - M_k\Phi^T M_{k+1}^{-1}(\hat{\underline{x}}_{k+1}^n - \hat{\underline{x}}_{k+1}^{k+1})$$

$$+ \sum_{j=1}^{n-k-1} M_k(\Phi^j)^T\Phi^T M_{k+j+1}^{-1}\hat{\underline{x}}_{k+j+1}^{k+j+1} - \sum_{j=1}^{n-k-1} M_k(\Phi^j)^T\Phi^T M_{k+j+1}^{-1}\hat{\underline{x}}_{k+j+1}^{k+j}$$

$$+ \sum_{j=1}^{n-k-1} M_k(\Phi^j)^T\Phi^T M_{k+j+1}^{-1}\hat{\underline{x}}_{k+j+1}^{k+j} - \sum_{j=1}^{n-k-1} M_k(\Phi^j)^T M_{k+j}^{-1}\hat{\underline{x}}_{k+j}^{k+j}$$

$$+ M_k\Phi^T M_{k+1}^{-1}\hat{\underline{x}}_{k+1}^n - M_k(\Phi^{n-k-1})^T\Phi^T M_n^{-1}\hat{\underline{x}}_n^n$$

$$= \underline{x}_k^k + P_k\Phi^T M_{k+1}^{-1} \cdot (\hat{\underline{x}}_{k+1}^n - \hat{\underline{x}}_{k+1}^k) - M_k\Phi^T M_{k+1}^{-1} \cdot (\hat{\underline{x}}_{k+1}^n - \hat{\underline{x}}_{k+1}^{k+1}) - M_k\Phi^T M_{k+1}^{-1}\hat{\underline{x}}_{k+1}^{k+1}$$

$$+ M_k(\Phi^{n-k-1})^T\Phi^T M_n^{-1}\hat{\underline{x}}_n^n + M_k\Phi^T M_{k+1}^{-1}\underline{x}_{k+1}^n - M_k(\Phi^{n-k-1})^T\Phi^T M_n^{-1}\hat{\underline{x}}_n^n$$

$$= \hat{\underline{x}}_k^k + P_k\Phi^T M_{k+1}^{-1}\cdot(\hat{\underline{x}}_{k+1}^n - \hat{\underline{x}}_{k+1}^k) .$$

This concludes the proof. |

5. A CONTINUITY PROPERTY OF STATE-PREDICTION DENSITIES

When both the $\underline{x}_k$ process and the error process v_k are Gaussian, the state-prediction density $f(\underline{x}_k|Y^{k-1})$ is Gaussian (since it is a posterior density for a completely Gaussian situation). However, the assumption that $f(\underline{x}_k|Y^{k-1})$ is Gaussian when v_k is non-Gaussian would seem highly dubious. In fact we have the following result.

PROPOSITION If both the state process $\underline{x}_k$ and the state prediction densities $f(\underline{x}_k|Y^{k-1})$, $k \geq 2$, are Gaussian, and ε_k has a (Gaussian) density f_ε, then the observation noise density is Gaussian.

Proof The state prediction density may be written

$$f(\underline{x}_k|Y^{k-1}) = \int f(\underline{x}_k|Y^{k-1},\underline{x}_{k-1})f(\underline{x}_{k-1}|Y^{k-1})d\underline{x}_{k-1}$$

$$= \int f_\varepsilon(\underline{x}_k - \Phi\underline{x}_{k-1})f(\underline{x}_{k-1}|Y^{k-1})d\underline{x}_{k-1} .$$

By a change of variables the above is recognized as a convolution integral. Since $f(\underline{x}_k|Y^{k-1})$, $k \geq 2$, is a Gaussian density obtained by convolving $f(\underline{x}_{k-1}|Y^{k-1})$ with the Gaussian density f_ε, it follows that $f(\underline{x}_{k-1}|Y^{k-1})$ is Gaussian for $k \geq 2$. Now

$$f(\underline{x}_{k-1}|Y^{k-1}) = c(Y^{k-1})f(\underline{x}_{k-1}|Y^{k-2})f_v(y_{k-1} - H\underline{x}_{k-1})$$

with $c(Y^{k-1}) = \{f(y_{k-1}|Y^{k-2})\}^{-1}$. Thus f_v is Gaussian. |

Although the above result rules out the possibility of $f(\underline{x}_k|Y^{k-1})$ being exactly Gaussian when f_v is non-Gaussian, it does not rule out the possibility of $f(\underline{x}_k|Y^{k-1})$ being nearly Gaussian when the same is true of f_v. Two hints at why the latter might be the case are as follows.

First of all, the state prediction density can be expressed as the (k-1)-fold convolution-like integral involving f_ε and f_v. Except in the special degenerate case where Φ is the identity matrix and $\sigma_\varepsilon^2 = 0$ so that $\underline{x}_k = \underline{x}$ is a constant, a normal convergence result will not hold. The reason is that the weights sequence in a corresponding weighted-sum representation tend to zero. None-the-less, the convolution-like form would lead one to expect central-limit-theorem-like effects to appear. Secondly, as was mentioned earlier, the Monte Carlo results in Masreliez (1975) and Martin and DeBow (1976) indicate that an ACM-filter based on the Gaussian assumption for $f(\underline{x}_k|Y^{k-1})$ will be nearly optimal.

The main result of this section is a continuity theorem which says that small changes in f_v from a Gaussian distribution will produce small changes in $f(\underline{x}_k|Y^{k-1})$

from a Gaussian distribution. The method with which we measure "small changes" in $f(\underline{x}_k|Y^{k-1})$ differs from our more conventional use of the Prohorov metric to measure "small changes" in f_v, and this difference is basic to the problem at hand. For the sake of notational simplicity the result will be stated for the special case of a first-order autoregression x_k. There is no difficulty in principle in extending the continuity result to vector $\underline{x}_k$ processes.

The state-prediction density $f(x_k|Y^{k-1})$ can be viewed as a function of the obser ation noise distribution F_v. The "value" of this functional at F_v is denoted $f(x_k|Y^{k-1};F_v)$.

The space of distributions $(\mathcal{F}_v,d_p)$ endowed with the Prohorov metric d_p is a complete metric space, and convergence in the Prohorov metric is equivalent to weak convergence (Billingsley, 1968).

The Prohorov metric has been an appealing one to use in robustness studies because it accomodates in a natural manner two important sources of deviation from a nominal distribution: (i) deviations which correspond to small changes in all of the data, and (ii) deviations which correspond to large changes in a small fraction of the data (Hampel, 1971). It may be shown for example that the contaminated normal distribution $CN(\gamma,\sigma_o^2,\sigma^2)$, which generates large outliers a fraction γ of the time on the average if $\sigma^2 >> \sigma_o^2$, is within Prohorov distance γ of the normal distribution $N(0,\sigma_o^2)$.

Suppose that $F_v^o = N(0,\sigma_o^2)$. What we would like to show is that when $d_p(F_v,F_v^o) < \delta$ the density $f(x_k|Y^{k-1};F_v)$ will be close to the Gaussian density $f(x_k|Y^{k-1};F_v^o) = N(x_k;\hat{x}_k^{k-1},M_k)$, provided $\delta > 0$ is small enough. Now we choose a distinctly different notion of the closeness than that associated with the Prohorov metric, with the following reasoning in mind. The Prohorov metric specifically allows a heavy-tailed distribution F_v to be close to F_v^o. What we now require is that $f(x_k|Y^{k-1};F_v)$ be close to $f(x_k|Y^{k-1};F_v^o)$ with regard to tail structure as well as overall location, scale and shape.

Thus we are led to consider $f(x_k|Y^{k-1};F_v)$ as being close to $f(x_k|Y^{k-1};F_v^o)$ if the score functions $\Psi_k(t;F_v) = -(\partial/\partial t)\ell og f(t|Y^{k-1};F_v)$ and $\Psi_k(t;F_v^o) = -(\partial/\partial t)\ell og f(t|Y^{k-1};F_v^o)$ are close in the sense that $|\Psi_k(t;F_v) - \Psi_k(t;F_v^o)|$ is small. Ideally we would like the above quantity to be small uniformly in t, $t \in R^1$, provided $d_p(F_v,F_v^o)$ is sufficiently small. The result to follow shows that this is true uniformily in t on arbitrary compact subsets C of the real line. Since $\Psi_k(t;F_v^o)$ defines a straight line for $F_v^o = N(0,\sigma_o^2)$, it will follow that $\Psi_k(t;F_v)$ is uniformly close to this straight line on an arbitrarily large interval $[-L,L]$.

The values $\Psi_k(t;F_v)$, $t \in C$ define a family of maps from $(\mathcal{F},d_p)$ to R^1, indexed by t. This family is denoted $\{\Psi_k(t;\cdot), t \in C\}$.

<u>THEOREM</u> (CONTINUITY OF STATE-PREDICTION DENSITIES) Suppose that the process $\underline{x}_k = x_k$ is a stationary Gaussian first-order autoregression with $\sigma_\varepsilon^2 = VAR\ \varepsilon_k$, transition parameter ϕ and that the i.i.d. sequence v_k has distribution F_v. Let C be a compact

subset of R^1 and let $k \geq 2$ be fixed. Then the family of maps $\{\Psi_k(t;\cdot),\ t \in C\}$ is equicontinuous at every F_v.

Proof The state-prediction density may be written in the form

$$f(x_k|Y^{k-1};F_v) = c(Y^{k-1})\int f(x_1)\prod_{i=1}^{k-1} f_\varepsilon(x_{i+1}-\phi x_i)d\tilde{F}_v(x_1,\ldots,x_{k-1})$$

where $c(Y^{k-1}) = \{f(Y^{k-1})\}^{-1}$ and

$$d\tilde{F}_v(x_1,\ldots,x_{k-1}) = \prod_{i=1}^{k-1} dF_v(y_i-x_i-\mu)$$

is the product measure obtained from the marginal measures $dF_v(y_i-x_i-\mu)$ with $y_1, \ldots, y_{k-1}$ fixed. Weak convergence of $\{F_{v,n}\}$ to F_v implies weak convergence of $\{\tilde{F}_{v,n}\}$ to $\tilde{F}_v$ (a consequence of Theorem 4.5 in Billingsley, 1968). Now

$$\Psi_k(t;F_v) = -(\partial/\partial x_k)\log f(x_k|Y^{k-1};F_v)\Big|_{x_k = t}$$

$$= -\frac{\int\left\{\frac{f'_\varepsilon(t-\phi x_{k-1})}{f_\varepsilon(t-\phi x_{k-1})}\right\}f_\varepsilon(t-\phi x_{k-1})\prod_{i=1}^{k-2} f_\varepsilon(x_{i+1}-\phi x_i)f(x_1)d\tilde{F}_v(x_1,\ldots,x_{k-1})}{\int f_\varepsilon(t-\phi x_{k-1})\prod_{i=1}^{k-2} f_\varepsilon(x_{i+1}-\phi x_i)f(x_1)d\tilde{F}_v(x_1,\ldots,x_{k-1})}$$

$$= \frac{1}{\sigma_\varepsilon^2}\,\frac{\int(t-\phi x_{k-1})f_\varepsilon(t-\phi x_{k-1})\prod_{i=1}^{k-2} f_\varepsilon(x_{i+1}-\phi x_i)f(x_1)d\tilde{F}_v(x_1,\ldots,x_{k-1})}{\int f_\varepsilon(t-\phi x_{k-1})\prod_{i=1}^{k-2} f_\varepsilon(x_{i+1}-\phi x_i)f(x_1)d\tilde{F}_v(x_1,\ldots,x_{k-1})}.$$

The integrand in the numerator defines a uniformly bounded family $\{g_t\}$ of functions on R^{k-1}, and this family is equicontinuous at each $x = (x_1, \ldots, x_{k-1}) \in R^{k-1}$. By Problem 8 on page 17 of Billingsley (1968)

$$\int g_t(x)\, d\tilde{F}_{v,n}(x) \to \int g_t(x)\, d\tilde{F}_v(x)$$

uniformly in t as $F_{v,n} \to F_v$ weakly, i.e., as $d_p(F_{v,n},F_v) \to 0$. The same property holds for the denominator integral with $\{g_t\}$ replaced by a slightly different family $\{\bar{g}_t\}$. Since the denominator is positive for $t \in C$ the result follows. |

Comment 1 The Gaussian assumption on x_1 and ε_k is not essential. The result of the theorem clearly holds for a broad class of non-Gaussian x_k processes, namely those for which the numerator integrand defines a family $\{g_t\}$ which is uniformly bounded and equicontinuous at each x.

Comment 2 The convergence of the numerator and denominator integrals is uniform for all $t \in R^1$. The compact set C is used only to avoid the problem caused by the fact that the denominator integral goes to zero as t goes to $\pm\infty$. It may be that a carefu analysis of the behavior of $\Psi_k(t;F_v)$ as $t \to \pm\infty$ will produce a stronger version of the theorem in which C is replaced by R^1.

Comment 3 The fact that the continuity property just established is rather special to the structure of a state prediction density $f(x_k|Y^{k-1})$ is revealed by comparison with two other cases. If $\Psi(t;F) = -(\partial/\partial t)\ell og f(t)$ is the score function (assuming it exists) for the distribution F, then it is easy to check that the mapping defined by $\Psi(t;F)$ is not continuous at any t. The second example is provided by the posterior density $f(x|y)$ for the scalar model $y = x + v$, with x Gaussian and independent of v. The score function of $f(x|y)$ does not enjoy the continuity property established in the above theorem - simple convolution of a non-Gaussian v with a Gaussian x clearly will not suffice.

6. CONCLUDING COMMENTS

Among the many important issues for further study, I would give priority to the following. Further experience in usage of ACM-smoothers on a wide variety of data sets is needed. The non-Gaussian nature of the state prediction density needs to be investigated in a quantitative manner. It needs to be determined whether or not filter and smoother correction terms which reflect potentially significant non-Gaussian aspects of the state prediction density are important, and if so, whether or not simple and useful correction terms exist.

REFERENCES

1. P. Billingsley (1968), Convergence of Probability Measures, John Wiley, New York NY.
2. W. S. Cleveland (1979), "Robust locally weighted regression and smoothing scatterplots," Jour. Amer. Statist. Assoc.
3. L. Denby and R. D. Martin (1979), "Robust estimation of the first-order autoregressive parameter," Jour. Amer. Stat. Assoc., 74, No. 365, 140-146.
4. F. Hampel (1971), "A general qualitative definition of robustness," Annals Math. Stat., 42, 1887-1896.
5. T. S. Huang, G. J. Yang and G. Y. Tang (1979), "A fast two-dimensional median filtering algorithm," IEEE Trans. on Acoustics, Speech and Signal Processing, 27, No. 1, 13-18.
6. P. Huber (1979), "Robust smoothing," in Robustness in Statistics, edited by R. L. Launer and G. Wilkinson, Academic Press, New York, NY.
7. A. Jazwinski (1970), Stochastic Processes and Filtering Theory, Academic Press, New York, NY.
8. B. Justusson (1977), "Statistical properties of median filters in signal and image processing" (unpublished report), Math. Inst., Royal Instit. of Tech., Stockholm, Sweden.

9. B. Kleiner, R. D. Martin and D. J. Thomson (1979), "Robust extimation of power spectra," Jour. Royal Statist. Soc. B, 41, No. 3.

10. R. V. Lenth (1977), "Robust splines," Comm. in Statistics-Theor. Methods, A6(9), 847-854.

11. C. L. Mallows (1979a), "Some theory of nonlinear smoothers," to appear in Annals of Statistics.

12. C. L. Mallows (1979b), "Resistant smoothing," to appear in Proc. Heidelberg Workshop on Smoothing Tech. for Curve Est. (T. Gasser and M. Rosenblatt, editors).

13. R. D. Martin (1979a), "Robust estimation of time series autoregressions," in Robustness in Statistics edited by R. L. Launer and G. Wilkinson, Academic Press.

14. R. D. Martin (1979b), "Robust methods for time-series," to appear in Proc. of International Time Series Meeting, Nottingham (O. D. Anderson, editor).

15. R. D. Martin (1979c), "Robust estimation of autoregressive models," in Time Series Analysis, Surveys and Recent Developments, Instit. of Math. Stat.

16. R. D. Martin and G. DeBow (1976), "Robust filtering with data-dependent covariance," Proc. of Johns Hopkins Conf. on Inf. Sciences and Systems.

17. R. D. Martin and J. E. Zeh (1979), "Robust generalized M-estimates for autoregressive parameters," Tech. Rep. No. 214, Dept. of Elec. Engr., Univ. of Wash., Seattle, WA (submitted to JASA).

18. C. J. Masreliez (1975), "Approximate non-Gaussian filtering with linear state and observation relations," IEEE Trans. on Auto Control, AC-20, 107-110.

19. J. S. Meditch (1967), "Orthogonal projection and discrete optimal linear smoothing," SIAM Jour. of Control, 5, No. 1, 74-89.

20. J. S. Meditch (1969), Stochastic Optimal Linear Estimation and Control, McGraw-Hill, New York, NY.

21. F. Mosteller and J. W. Tukey (1977b), Data Analysis and Regression, Addison-Wesley, Reading, MA.

22. L. R. Rabiner, M. R. Sambur and C. E. Schmidt (1975), "Applications of a nonlinear smoothing algorithm to speech processing," IEEE Trans. Acoust., Speech, Signal Proc., ASSP-23, 552-557.

23. B. W. Stuck (1976), "Minimum error dispersion linear filtering of symmetric stable processes," Bell Laboratories Tech. Memo., Murray Hill, NJ.

24. W. Stuetzle (1979), "Asymptotics for running M-estimates," to appear in Proc. of Heidelberg Workshop on Smoothing Tech. for Curve Est. (T. Gasser and M. Rosenblatt, editors).

25. D. J. Thomson (1977), "Spectrum estimation techniques for characterization and development of WT4 waveguide-I," Bell System Tech. Jour., 56, No. 4, 1769-1815.

26. J. W. Tukey (1977a), Exploratory Data Analysis, Addison-Wesley, Reading, MA.

27. P. Velleman (1975), "Robust nonlinear data smoothing," Tech. Rep. No. 89, Series 2, Dept. of Statistics, Princeton Univ.

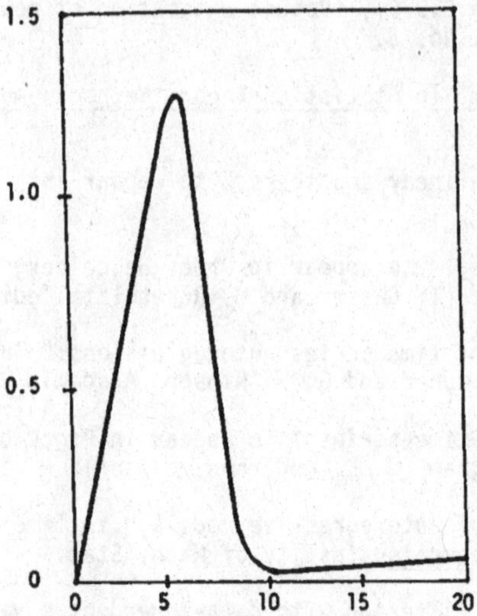

Figure 1a. Shape of $\Psi_k(\tilde{y}_k)$ for $\gamma = .02$, $\sigma_{ok} = 2$, $\sigma_k = 20$.

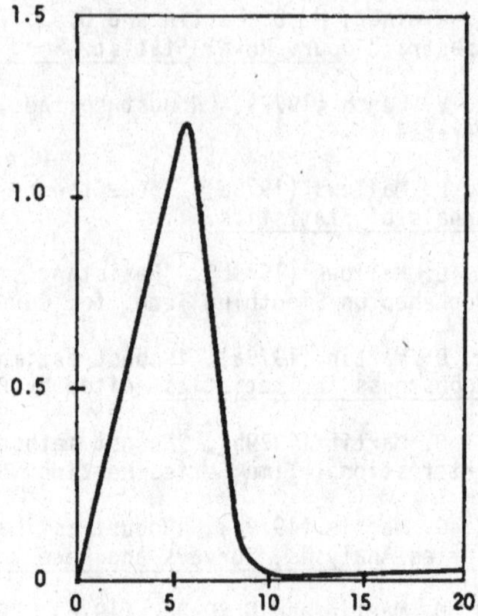

Figure 1b. Shape of $\Psi_k(\tilde{y}_k)$ for $\gamma = .1$, $\sigma_{ok} = 2$, $\sigma_k = 60$.

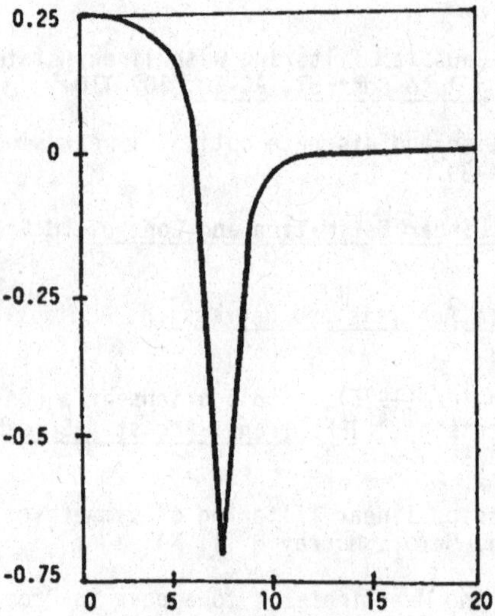

Figure 1c. Shape of $\Psi_k'(\tilde{y}_k)$ for $\gamma = .02$, $\sigma_{ok} = 2$, $\sigma_k = 20$.

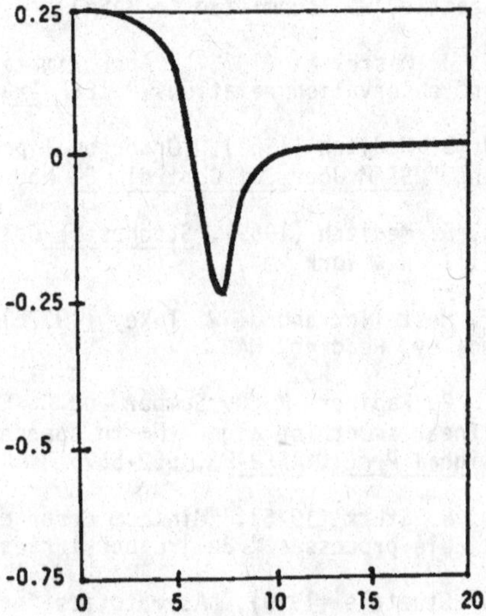

Figure 1d. Shape of $\Psi_k'(\tilde{y}_k)$ for $\gamma = .1$, $\sigma_{ok} = 2$, $\sigma_k = 60$.

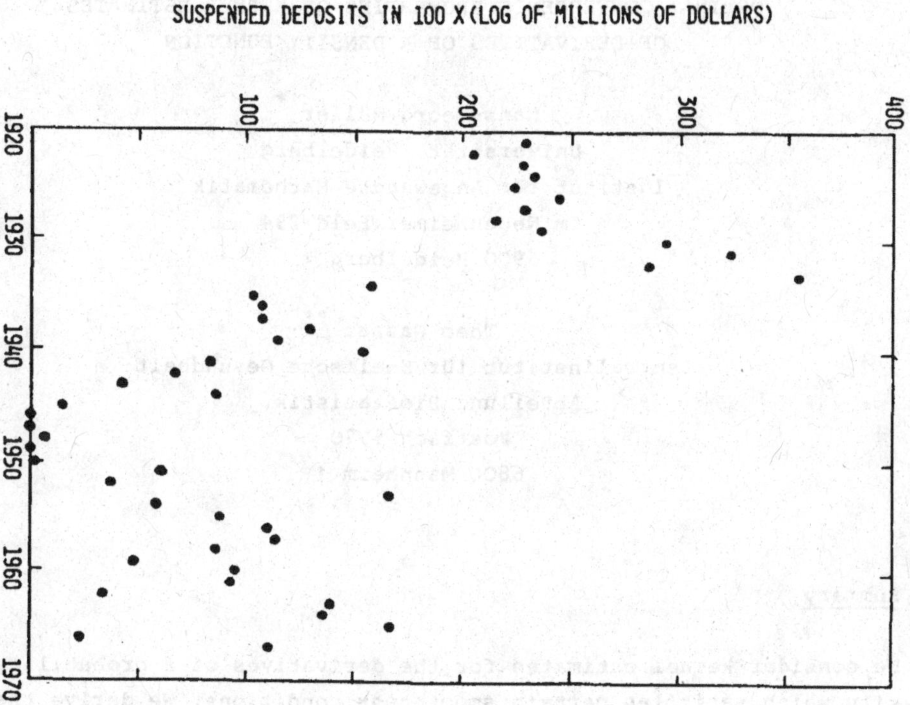

FIGURE 2A. SUSPENDED BANK DEPOSITS DATA

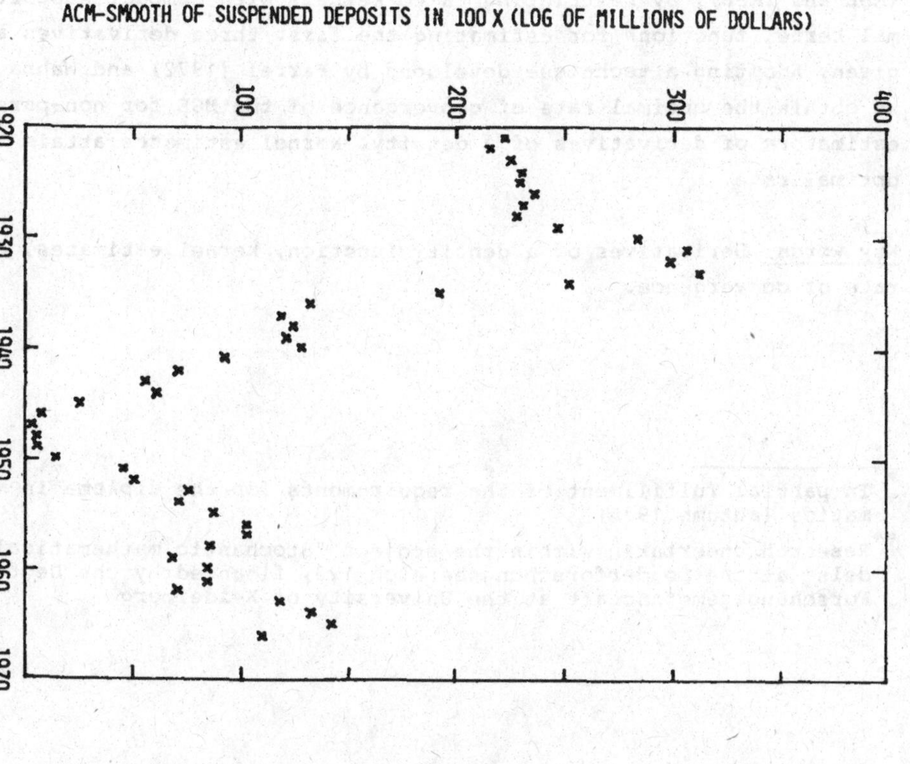

FIGURE 2B. ACM-SMOOTH OF SUSPENDED BANK DEPOSITS DATA

OPTIMAL CONVERGENCE PROPERTIES OF KERNEL ESTIMATES OF DERIVATIVES OF A DENSITY FUNCTION

Hans-Georg Müller *
Universität Heidelberg
Institut für Angewandte Mathematik
Im Neuenheimer Feld 294
6900 Heidelberg 1

Theo Gasser **
Zentralinstitut für Seelische Gesundheit
Abteilung Biostatistik
Postfach 5970
6800 Mannheim 1

Summary

We consider kernel estimates for the derivatives of a probability den- sity which satisfies certain smoothness conditions. We derive the rat of convergence of the local and of the integrated mean square error (MSE and IMSE), by restricting us to kernels with compact support. Op- mal kernel functions for estimating the first three derivatives are given. Adopting a technique developed by Farrel (1972) and Wahba (197 we obtain the optimal rate of convergence of the MSE for non-parametr estimators of derivatives of a density. Kernel estimates attain this optimal rate.

Key words: Derivatives of a density function, kernel estimates, optim rate of convergence.

---

* In partial fulfillment of the requirements for the diploma in mathe matics (autumn 1978)

** Research undertaken within the project "Stochastic mathematical mo- dels" at the Sonderforschungsbereich 123, financed by the Deutsche Forschungsgemeinschaft at the University of Heidelberg

. Asymptotic Properties of Kernel Estimates

et $X_1, \dots, X_n$ be i.i.d. real-valued random variables with density f . A ernel estimate for f is defined as follows:

$$f_n(x) = \frac{1}{nb(n)} \sum_{j=1}^{n} w_o \left(\frac{x - X_j}{b(n)}\right)$$

his estimate was introduced by Rosenblatt (1956) and further investiated by Parzen (1962). The kernel function w_o is assumed to be interable, and b(n) is a sequence of non-negative real bandwidths. Throughut this paper we restrict ourselves to kernel functions with compact upport. This is justified for a number of reasons: It will turn out hat kernels with compact support suffice to attain the optimal rate f convergence of MSE; any kernel with non-compact support has to be runcated in practice. Finally, this assumption simplifies the proofs.

f w_o is bounded with $\int w_o(v)\,dv = 1$, and if f is bounded and f is coninuous in x_o, then $f_n(x_o)$ is a MSE-consistent estimate of $f(x_o)$, if

$$\lim_{n\to\infty} b(n) = 0, \lim_{n\to\infty} nb(n) = \infty \text{(Rosenblatt 1971)}.$$

ow let us assume ν, $k \in \mathbb{N}$, $\nu \geq 0$, $\nu + 2 \leq k$ and f ν times differentiable. s a kernel estimate for $f^{(\nu)}$ we define

$$f_{n,\nu}(x) = \frac{1}{nb^{\nu+1}(n)} \sum_{j=1}^{n} w_\nu \left(\frac{x - X_j}{b(n)}\right)$$

here w_ν is derived from a ν times differentiable function $w: \mathbb{R} \to \mathbb{R}$ with

1) $w^{(\nu)} = w_\nu$, Support $w = [-\tau, \tau]$,

$$\int_{-\tau}^{\tau} w(v)\,dv = 1, \quad w^{(j)}(\tau) = w^{(j)}(-\tau) = 0 \, (j = 0 \dots \nu-1).$$

he computation of the variance of this estimate is straightforward, and or the expectation we use (1) and an iterated partial integration:

2) $$E\,(f_{n,\nu}(x)) = \frac{1}{b^{\nu}(n)} \int_{-\tau}^{\tau} w_\nu\,(v)\; f(x-vb(n))\,dv$$

$$= \int_{-\tau}^{\tau} w(v)\ f^{(\nu)}(x-vb(n))\ dv$$

(3) $$\mathrm{Var}\ (f_{n,\nu}\ (x)) = \frac{1}{nb^{2\nu+1}(n)} \int_{-\tau}^{\tau} w_\nu^{\ 2}(v) f(x-vb(n))\ dv$$

$$- \frac{1}{nb^{2\nu}(n)} \left(\int_{-\tau}^{\tau} w_\nu(v) f(x-vb(n))\ dv\right)^2 .$$

Our estimate is thus consistent in MSE if:

(4) $\lim_{n\to\infty} b(n) = 0,\quad \lim_{n\to\infty} nb^{2\nu+1}(n) = \infty$ and

$f^{(\nu)}$ is continuous in x.

The following theorem gives the rate of convergence for the IMSE and the MSE, as well as the dependence on the kernel functions, evaluated for the optimal bandwidth.

Theorem 1 Let f be k times differentiable and the kernel function w_ν satisfy the following conditions:

(5) $$\begin{cases} \text{Support } w_\nu = [-\tau, \tau] \\ \int_{-\tau}^{\tau} w_\nu\ (v) v^j\ dv = \begin{cases} 0 & j = 0 \ldots \nu-1,\ \nu+1 \ldots k-1 \\ (-1)^\nu\ \nu\,! & j = \nu \end{cases} \\ \text{and } \int_{-\tau}^{\tau} w_\nu\ (v)\ v^k\ dv \neq 0. \end{cases}$$

Let $C := C_0\ \left[\int_{-\tau}^{\tau} w_\nu^{\ 2}\ (v)\ dv\right]^{\frac{2(k-\nu)}{2k+1}} \left[\int_{-\tau}^{\tau} w_\nu\ (v)\ v^k\ dv\right]^{2\cdot\frac{2\nu+1}{2k+1}}$

(C_0 depending on ν, k only).

A. If f is uniformly bounded, f, $f^{(k)} \in L^2$, $f^{(k)}$ continuous and

$\int f^{(k)^2}(x)\ dx \neq 0$, then

(6) $$E \int_{-\infty}^{\infty} (f_{n,\nu}(x) - f^{(\nu)}(x))^2 dx = n^{-\frac{2(k-\nu)}{2k+1}} \left[\int_{-\infty}^{\infty} f^{(k)^2}(x) dx\right]^{\frac{2\nu+1}{2k+1}} \cdot C + o\left(n^{-\frac{2(k-\nu)}{2k+1}}\right)$$

B. If $f^{(k)}$ is continuous in x_0, $f^{(k)}(x_0)\ f\ (x_0) \neq 0$, then

$$(7)\quad E(f_{n,\nu}(x_o)-f^{(\nu)}(x_o))^2 = n^{-\frac{2(k-\nu)}{2k+1}} f^{(k)}(x_o)^{2\cdot\frac{2\nu+1}{2k+1}} f(x_o)^{\frac{2(k-\nu)}{2k+1}} \cdot C + o\left(n^{-\frac{2(k-\nu)}{2k+1}}\right)$$

Remark: If $\int f^{(k)^2}(x)dx = 0$, or $f^{(k)}(x_o)f(x_o) = 0$ respectively, the rate $n^{-\frac{2(k-\nu)}{2k+1}}$ is also obtained.

Proof: First note that (5) is equivalent to (1) and

$$\int_{-\tau}^{\tau} w(v)v^j\,dv = \begin{cases} 1 & j=0 \\ 0 & j=1\ldots k-\nu-1 \\ \neq 0 & j=k-\nu \end{cases}$$

This can be seen by partial integration . From (3) we conclude, using the boundedness of f and the compactness of support of w_ν ,

$$(8)\quad \int_{-\infty}^{\infty} \mathrm{Var}(f_{n,\nu}(x))dx = \frac{1}{nb^{2\nu+1}(n)} \int_{-\tau}^{\tau} w_\nu^2(v)dv + O\left(\frac{1}{nb^{2\nu}(n)}\right)$$

By a Taylor expansion of $f^{(\nu)}$, using (2) and the orthogonality conditions (5), we obtain for the bias ($\Theta_v \in [0,1]$):

$$\int_{-\infty}^{\infty} (E f_{n,\nu}(x) - f^{(\nu)}(x))^2\,dx$$

$$= \frac{1}{(k-\nu)!^2} b^{2(k-\nu)}(n) \int_{-\infty}^{\infty} \left[\int_{-\tau}^{\tau} w(v)v^{k-\nu} f^{(k)}(x-\Theta_v\, b(n)\, v)\,dv\right]^2 dx$$

$$= \frac{1}{(k-\nu)!^2} b^{2(k-\nu)}(n) \int_{-\infty}^{\infty} f^{(k)^2}(x)\,dx\left[\int_{-\tau}^{\tau} w(v)v^{k-\nu}\,dv\right]^2 + o(b^{2(k-\nu)}(n))$$

The last step follows from

$$\int_{-\infty}^{\infty} \left[\int_{-\tau}^{\tau} w(v)\, v^{k-\nu}\Delta(x,v,n,k)\,dv\right]^2 dx \to 0 \quad (n\to\infty)$$

$$\int_{-\infty}^{\infty} \left[\int_{-\tau}^{\tau} w(v)v^{k-\nu} f^{(k)}(x)dv\right]\left[\int_{-\tau}^{\tau} w(v)v^{k-\nu}\Delta(x,v,n,k)dv\right] dx \to 0 \quad (n\to\infty)$$

where: $f^{(k)}(x-\Theta_v\, b(n)v) =: f^{(k)}(x) + \Delta(x,v,n,k)$

by the compactness of support of w, the continuity of $f^{(k)}$, $f^{(k)} \in L^2$ and several applications of the Cauchy-Schwarz inequality (for details see Müller (1978), Theorem 3). By partial integration we get

$$\frac{1}{(k-\nu)!^2} [\int_{-\tau}^{\tau} w(v) v^{k-\nu} dv]^2 = \frac{1}{k!^2} [\int_{-\tau}^{\tau} w_\nu (v) v^k dv]^2 ,$$

thus we have

$$\text{IMSE} = \frac{1}{nb^{2\nu+1}(n)} \int_{-\tau}^{\tau} w_\nu^2(v) dv +$$

$$\frac{1}{k!^2} b^{2(k-\nu)}(n) \int_{-\infty}^{\infty} f^{(k)^2}(x)dx \, [\int_{-\tau}^{\tau} w_\nu(v)v^k dv]^2 + o\left(\frac{1}{nb^{2\nu+1}(n)} + b^{2(k-\nu)}(n)\right)$$

By inserting the optimal value of b(n), which minimizes this expression we obtain the assertion of part A of the theorem. The remark is clear by simply choosing $b(n) = n^{-\frac{1}{2k+1}}$. The proof of B needs the same kind of reasoning.

A result similar to Theorem 1 B has also been obtained by Singh (1977).

There is an interesting application of Theorem 1 to the problem of finding optimal kernels, i.e. functions which minimize the expression:

$$(\int_{-\tau}^{\tau} w_\nu^2(v)dv)^{k-\nu} \left| \int_{-\tau}^{\tau} w_\nu(v) v^k dv \right|^{2\nu+1}$$

(where τ is not fixed) and satisfy the conditions given in (5). This problem is not well posed without further assumptions. Epanechnikov (1969) gives the optimal kernel for $\nu = 0$, $k = 2$ with the additional requirement of nonnegativity and symmetry. It is plausible to ask for a symmetric kernel if ν is even and for an antisymmetric one if ν is odd. A kernel which satisfies (5) must have at least (k-2) changes of sign. It is called minimal in the class of kernels of order k, if it has exactly (k-2) changes of sign (compare Gasser & Müller, this volume). With the requirement of minimality we can compute the optimal kernels for the first three derivatives ($k = \nu + 2$). These kernels are polynomials of degree k, restricted to the interval with their outermost zeroes as endpoints. They are given numerically in the following table, normalized to the interval [-1, 1], and their graphs are displayed in fig. 1.

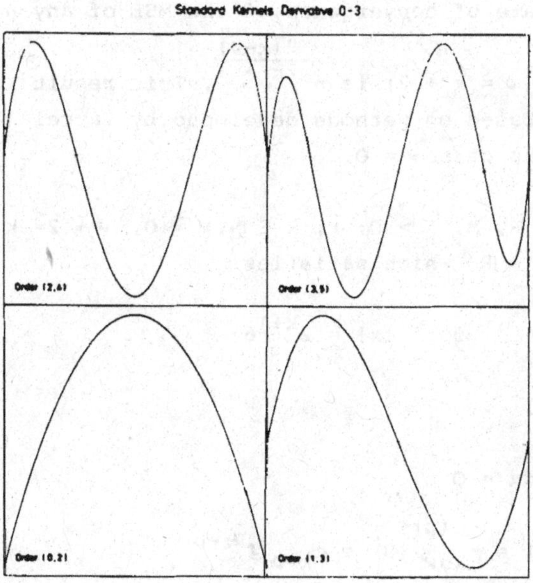

fig. 1: Plots of the optimal kernels of order (ν,k), $\nu = 0\text{-}3$, $k = \nu + 2$

TABLE: Optimal kernels for estimating derivatives (normalized to [-1,1])

ν	k	kernel function	$\beta = \int_{-1}^{1} w_\nu(v) v^k dv$
0	2	$0.75 - 0.75\, v^2$	0.2000
1	3	$-3.75\, v + 3.75\, v^3$	-0.4286
2	4	$-6.5625 + 39.3750\, v^2 - 32.8125\, v^4$	1.333
3	5	$177.2\, v - 590.6\, v^3 + 413{,}4\, v^5$	-5.455

2. Optimal Rate of Convergence

If the densities considered are contained in

$$(10)\quad S_k(M) = \left\{ f: \mathbb{R} \to \mathbb{R} \mid f \;\; k \text{ times continuously differentiable}, \; \sup_{x \in \mathbb{R}} |f^{(k)}(x)| < M \right\},$$

then the optimal rate of convergence of the MSE of any nonparametric estimator of $f^{(\nu)}$ $(\nu \leq k+2)$ is $n^{-\frac{2(k-\nu)}{2k+1}}$. This result is given in Theorem 2, and is based on methods developed by Farrel (1972) and Wahba (1975) for the case $\nu = 0$.

Lemma 1 For $M, \delta \in \mathbb{R}$, $M, \delta > 0$; $\nu, k \in \mathbb{N}$, $\nu \geq 0$, $\nu + 2 \leq k$, there is a function $\psi_{k\delta\nu} : \mathbb{R} \to \mathbb{R}$ which satisfies:

(i) $\psi_{k\delta\nu}(x) = 0 \quad$ for $|x| \geq 2^{k+1}\delta$

(ii) $\psi_{k\delta\nu} \in S_k(M)$

(iii) $\int_{-\infty}^{\infty} \psi_{k\delta\nu}(x)\,dx = 0$

(iv) $\max_x |\psi_{k\delta\nu}^{(\nu)}(x)| = \psi_{k\delta\nu}^{(\nu)}(0) = c_{k-\nu}\delta^{k-\nu}$,

$$\max_x |\psi_{k\delta\nu}(x)| = c_k \delta^k$$

(v) $\int_{-\infty}^{+\infty} \psi_{k\delta\nu}^2(x)\,dx \leq d_k \delta^{2k+1}$

where $c_{k-\nu}$, d_k are constants.

Proof: We define a function $\varphi_{k\delta}$ recursively:

$$\varphi_{0\delta}(x) := \begin{cases} \frac{2M}{\delta}x + M & -\frac{\delta}{2} \leq x \leq 0 \\ -\frac{2M}{\delta}x + M & 0 \leq x \leq \frac{\delta}{2} \\ 0 & \text{elsewhere} \end{cases}$$

$$\varphi_{j+1,\delta}(x) := \int_{-\infty}^{x} (\varphi_{j\delta}(y + 2^{j-1}\delta) - \varphi_{j\delta}(y - 2^{j-1}\delta))\,dy$$

The following relations are established by induction:

(i') $\varphi_{k\delta}(x) = 0 \quad$ for $|x| \geq 2^{k-1}\delta$

(ii') $\varphi_{k\delta} \in S_k(M)$

(iii') $\max_x |\varphi_{k\delta}(x)| = \varphi_{k\delta}(0) = c_k\delta^k$ with constants c_k

(iv') $\max_x |\varphi_{k\delta}^{(\nu)}(x)| = \varphi_{k\delta}^{(\nu)}\,[-\delta[2^{k-2} + 2^{k-3} + \ldots + 2^{k-(\nu+1)}]$

$$= c_{k-\nu}\,\delta^{k-\nu} \text{ for } \nu \geq 1,\ \nu + 2 \leq k$$

Define

$$\psi_{k\delta\nu}(x) := \varphi_{k\delta}\,(x-\delta\,(2^{k-2} + 2^{k-3} + \ldots + 2^{k-(\nu+1)})$$

$$-\varphi_{k\delta}\,(x-\delta\,(-2^{k} + 2^{k-2} + \ldots + 2^{k-(\nu+1)})$$

(For $\nu = 0$, we set $2^{k-2} + 2^{k-3} + \ldots + 2^{k-(\nu+1)} = 0$)

Thus $\psi_{k\delta\nu}$ is composed of two "blocks" of functions $\varphi_{k\delta}$ in such a way that the absolute maximum of the ν-th derivative is attained at 0. The relations (i) - (v) follow from (i') - (iv') (details are given in Müller (1978) Lemmata 11 and 12).

Let P_f be the probability measure belonging to the density f, $\hat{f}_{n,\nu}(X_1 \ldots X_n)(x_o)$ an estimator of $f^{(\nu)}(x_o)$ based on the i.i.d. random variables $X_1 \ldots X_n$ with density f.

<u>Theorem 2</u> Assume ν, $k \in \mathbb{N}$, $k \geq \nu + 2$, $L > 0$ an arbitrary real number, α_n a nonnegative real-valued sequence.

If $\sup_{f \in S_k(L)} E_f\,(\hat{f}_{n,\nu}(x_o) - f^{(\nu)}(x_o))^2 = \alpha_n\, n^{-\frac{2(k-\nu)}{2k+1}}$,

then $\liminf_{f \in S_k(L)} \alpha_n > 0$, i.e. there is an $\varepsilon > 0$

and $n_o \in \mathbb{N}$ such that $\alpha_n > \varepsilon$ for all $n > n_o$.

<u>Proof:</u> For any nonnegative sequence β_n define

$$\gamma_n := \inf_{f \in S_k(L)} P_f\,(\,|\hat{f}_{n,\nu}(x_o) - f^{(\nu)}(x_o)| \leq \beta_n)$$

<u>Lemma 2</u>

(11) If $\lim_{n\to\infty} \gamma_n = 1$, then $\lim_{n\to\infty} n^{\frac{2(k-\nu)}{2k+1}}\,\beta_n^{\,2} = \infty$.

<u>Proof</u> of this Lemma: From $\lim_{n\to\infty} \gamma_n = 1$, we have

$$\varepsilon_n := \frac{n}{d_k}\,[(\frac{\gamma_n^2}{1-\gamma_n})^{\frac{1}{n}} - 1] \to \infty$$

(where d_k is the constant occurring in Lemma 1 part (v)). We may choose a real-valued sequence δ_n such that $0 < \delta_n < \varepsilon_n$ and $\delta_n \to \infty$, $\frac{\delta_n}{n} \to 0$ $(n\to\infty)$. Let us define $\delta(n) := (\frac{\delta n}{n})^{\frac{1}{2k+1}}$.

Then we obtain

$$(12)\quad \gamma_n + \frac{\gamma_n^2}{(1+d_k\,\delta^{2k+1}(n))^n} > 1.$$

Let f_1 be a density with $f_1(x) > 0$ $(x\in\mathbb{R})$, $f_1\in S_k(\frac{L}{2})$, $f_1(x) = 1$ in some neighbourhood of x_o. As $\delta(n) \to 0$ $(n\to\infty)$, there exists a n_o such that for $n > n_o$ we have:

$$(13)\quad 1 > c_k\,\delta(n)^k \text{ and } f_1(x) = 1, \text{ if } |x-x_o| < 2^{k+1}\delta(n).$$

Only $n>n_o$ will be considered in what follows.
Let $\Psi_{k\delta(n)\nu}$ be a sequence of functions with the properties (i) - (v) of the first Lemma, where

$M = \frac{L}{2}$. Define $f_{2,n}(x) := f_1(x) + \Psi_{k\delta(n)\nu}(x-x_o)$.

We conclude:

$f_{2,n} \in S_k(L)$, from (ii)

$f_{2,n}$ is a density, from (i), (iii), (iv), (13).

From (13), (iii), (iv) we obtain the following inequality,

$$\int_{-\infty}^{\infty}\cdots\int_{-\infty}^{\infty}\prod_{i=1}^{n}\frac{f_{2,n}^2(x_i)}{f_1(x_i)}\,dx_i \leq [1 + d_k\delta(n)^{2k+1}]^n.$$

This relation, together with the inequality of Cauchy-Schwarz, yields

$$P_{f_2}(|\hat{f}_{n,\nu}(X_1,\dots,X_n)(x_o) - f_{2,n}^{(\nu)}(x_o)| \leq \beta_n)$$

$$\leq P_{f_1}(|\hat{f}_{n,\nu}(X_1,\dots,X_n)(x_o) - f_{2,n}^{(\nu)}(x_o)| \leq \beta_n)^{\frac{1}{2}}(1 + d_k\delta(n)^{2k+1})^{\frac{n}{2}}.$$

This implies $\frac{\gamma_n^2}{(1+d_k\delta(n)^{2k+1})^n} \leq P_{f_1}(|\hat{f}_{n,\nu}(x_o) - f_{2,n}^{(\nu)}(x_o)| \leq \beta_n)$.

Combining this result with (12) yields

$$P_{f_1}(|\hat{f}_{n,\nu}(x_o) - f_1^{(\nu)}(x_o)| \leq \beta_n) + P_{f_1}(|\hat{f}_{n,\nu}(x_o) - f_{2,n}^{(\nu)}(x_o)| \leq \beta_n) > 1.$$

When we relate this to (iv), we get the assertion of the Lemma:

$$c_{k-\nu}\,\delta(n)^{k-\nu} = |\Psi_{k\delta(n)\nu}^{(\nu)}(0)| = |f_1^{(\nu)}(x_o) - f_{2,n}^{(\nu)}(x_o)| \leq 2\,\beta_n$$

and thus

$$n^{\frac{2(k-\nu)}{2k+1}}\,\beta_n^{\,2} \geq \frac{c_{k-\nu}^{\,2}}{4}\,\delta_n^{\frac{2(k-\nu)}{2k+1}} \to \infty.$$

We choose now $\beta_n = n^{-\frac{(k-\nu)}{2k+1}}$ and apply the Čebyšev inequality:

$$\alpha_n = \frac{1}{\beta_n^2}\sup_{f\in S_k(L)} E_f(\hat{f}_{n,\nu}(x_o) - f^{(\nu)}(x_o))^2 \geq 1 - \gamma_n.$$

Observing $\lim_{n\to\infty} n^{\frac{2(k-\nu)}{2k+1}}\beta_n^{\,2} = 1$ we conclude by (11)

$$\liminf_{n\to\infty} \alpha_n \geq \liminf_{n\to\infty} (1-\gamma_n) > 0$$

which completes the proof.

From (10) we may conclude especially

$$\sup_{f\in S_k(L)}\ \sup_{x\in\mathbb{R}} f(x) < K < \infty \text{ (for some constant K).}$$

Thus from Theorem 1B we have for the kernel estimate $f_{n,\nu}$

$$\sup_{f\in S_k(L)} E_f(f_{n,\nu}(x_o) - f^{(\nu)}(x_o))^2 = n^{-\frac{2(k-\nu)}{2k+1}}\,L^{2\frac{2\nu+1}{2k+1}}\,K^{\frac{2(k-\nu)}{2k+1}}\cdot C + o(n^{-\frac{2(k-\nu)}{2k+1}})$$

The kernel estimate $f_{n,\nu}$ discussed here has therefore optimal convergence properties.

References

Epanechnikov, V.A. (1969).Nonparametric estimation of a multivariate probability density. Theor. Prob. Appl. 14, 153-158

Farrel, R.H. (1972). On the best obtainable asymptotic rates of convergence in estimation of a density function at a point. Ann. Math. Statist. 43, 170-180

Müller, H.G. (1978). Kernschätzer für Wahrscheinlichkeitsdichten und Regressionsfunktionen. Diplomarbeit Universität Heidelberg, Math. Fak.

Parzen, E. (1962). On estimation of a probability density function and mode. Ann. Math. Statist. 35, 1065-1076

Rosenblatt, M. (1956). Remarks on some nonparametric estimates of a density function. Ann. Math. Statist. 27, 832-835

Rosenblatt, M. (1971). Curve estimates. Ann. Math. Statist. 42, 1815-1841

Singh, R.S. (1977). Improvement on some known nonparametric uniformly consistent estimators of derivatives of a density. Ann. Statist. 5, 394-399

Wahba, G. (1975). Optimal convergence properties of variable knot, kernel and orthogonal series methods for density estimation. Ann. Statist. 3, 15-29

DENSITY QUANTILE ESTIMATION APPROACH TO STATISTICAL DATA MODELLING

by

Emanuel Parzen

Institute of Statistics

Texas A&M University

Abstract

This paper describes the density-quantile function approach to statistical analysis of a sample as involving five phases requiring the study of various population raw and smoothed quantile and density-quantile functions. The phases can be succinctly described in terms of the notation for the functions studied: (i) Q, fQ, q, (ii) $\tilde{Q}$, $\tilde{q}$, (iii) $\bar{f}Q$, (iv) $\hat{f}Q$, $\hat{d}$, $d(u) = f_0Q_0(u)q(u)/\sigma_0$, $\sigma_0 = \int_0^1 f_0Q_0(u)q(u)du$, (v) $\hat{Q} = \hat{\mu} + \hat{\sigma}Q_0$.

Research supported by grant DAAG29-78-G-0180 from the Army Research Office.

0. Introduction

The density-quantile function approach to modeling a statistical data set consisting of a sample of observations of a random variable X regards the process of statistical data analysis as involving five phases.

(1) Probability based data analysis. When asked the exploratory question "here is a data set; what can be concluded," what we desire to draw conclusions about is the probability distribution from which the sample purports to be a representative sample. Standard statistical theory is concerned with inferring from a sample the properties of a random variable that are expressed by its distribution function $F(x) = Pr[X \leq x]$ and its density function $f(x) = F'(x)$. I propose that greater insight will be obtained by formulating conclusions in terms of the qualitative and quantitative behavior of the quantile function $Q(u) = F^{-1}(u)$, $0 \leq u \leq 1$, and the density-quantile function $fQ(u) = f(Q(u))$, $0 \leq u \leq 1$. We should become familiar with the possible shapes these functions could have.

(2) Sample Quantile Function. Much current statistical theory is concerned with the properties of a sample that can be expressed in terms of its sample distribution function $\tilde{F}(x)$, $-\infty < x < \infty$, defined by

$$\tilde{F}(x) = \text{proportion of the sample with values} \leq x .$$

I propose that the basic descriptive statistics of a sample is its sample quantile function $\tilde{Q}(u)$ defined so that it has a derivative $\tilde{q}(u) = \tilde{Q}'(u)$, called the sample quantile-density function. Exploring the data for patterns, as well as modelling the data, consists of examining how well various theoretical quantile functions $Q(u)$ match, or fit, $\tilde{Q}$.

(3) Sample Density-Quantile Function $\tilde{fQ}$. The most widely used graphical procedure for inspecting a sample is the histogram. I propose as an alternative a raw estimator $\tilde{fQ}$ of fQ, which can be obtained in several ways. The graph of $\tilde{fQ}$ provides insights into the type of distribution which the data possesses, including the following types: Symmetric; J-shaped; Skewed to the right; Skewed to the left; Uniform; Normal; Exponential; Short-tailed (limited type); Long-tailed (Cauchy or Student t type); Exponential-tailed (Weibull type); Bimodal (or multimodal); Zeroes in density; Outliers; Discrete (infinities in density).

(4) Smoothed Density-Quantile Function $\hat{fQ}$. The qualitative behavior or shape of the density-quantile function fQ classifies the type of probability distribution which a random variable possesses. To answer estimation questions about a random variable, we need quantitative estimators $\hat{fQ}$ of fQ which can be accomplished by a variety of smoothing or density estimation methods. I propose that an easily implementable procedure is provided by autoregressive smoothing of $f_0Q_0(u)\tilde{q}(u)$, where $f_0Q_0(u)$ is a "flattening" function specified by the statistician. This approach also provides goodness-of-fit tests of the hypothesis

$$H_0: F(x) = F_0(\frac{x-\mu}{\sigma}) \ , \quad Q(u) = \mu + \sigma Q_0(u)$$

where F_0 is a specified distribution function with quantile function $Q_0(u)$, and μ and σ are location and scale parameters to be estimated.

(5) Parametrically Smoothed Quantile Function $\hat{Q}$. To complete the process of modeling the sample quantile function $\tilde{Q}$ (that is, fitting

$\tilde{Q}$ by a theoretical quantile function Q), one postulates a parametric model such as $Q(u) = \mu + \sigma Q_0(u)$ to be fitted to $\tilde{Q}(u)$. Parameters such as μ and σ can be efficiently estimated by regression analysis of the continuous parameter "time series" $\tilde{Q}(u)$ using the theorem that the asymptotic distribution of $fQ(u)\{\tilde{Q}(u) - Q(u)\}$ is a Brownian bridge stochastic process. A final check of the model is provided by the goodness of fit of $\hat{Q}(u) = \hat{\mu} + \hat{\sigma} Q_0(u)$ to $\tilde{Q}(u)$.

Some of the details involved in carrying out the foregoing phases of statistical data modeling are described in this paper.

1. Quantile Functions and Density Quantile Functions

Corresponding to a distribution function $F(x)$, $-\infty < x < \infty$, we define its quantile function $Q(u)$, $0 \leq u \leq 1$, to be its inverse:

$$Q(u) = F^{-1}(u) = \inf\{x: F(x) \geq u\} . \tag{1}$$

Note that we use u to denote the argument of Q; it is a variable in the unit interval: $0 \leq u \leq 1$.

Two identities are so useful that I give them names:

Correspondence Identity: $F(x) \geq u$ if, and only if, $Q(u) \leq x$; (2)

Inverse Identity: $FQ(u) = u$ if F is continuous . (3)

Three important functions are defined by

Quantile density function $q(u) = Q'(u)$ (4)

Density-quantile function $fQ(u) = f(Q(u))$ (5)

Score function $J(u) = -(fQ)'(u)$. (6)

The shapes of these functions turn out to be independent of location and scale parameters.

By differentiating the Inverse Identity, we obtain

$$\text{Reciprocal Identity:} \quad fQ(u)q(u) = 1 . \tag{7}$$

In words, the quantile-density and density-quantile function are reciprocals of each other.

One important consequence of the Reciprocal Identity is the agreement of our definition of the score function J(u) (in terms of the derivative of fQ(u)) with the definition given in the theory of non-parametric statistical inferences

$$J(u) = \frac{-f'(F^{-1}(u))}{f(F^{-1}(u))} . \tag{8}$$

It seems easier to estimate J(u) using formula (6) rather than formula (8).

In the density-quantile function approach a basic role is played by the density-quantile function of the normal distribution

$$\Phi(x) = \int_{-\infty}^{x} \phi(y)dy , \quad \phi(y) = \frac{1}{\sqrt{2\pi}} \exp(-\tfrac{1}{2} y^2) .$$

The quantile function $\Phi^{-1}(u)$ has to be computed numerically. Then one computes the density-quantile function by

$$\phi\Phi^{-1}(u) = \frac{1}{\sqrt{2\pi}} \exp -\tfrac{1}{2}\{\Phi^{-1}(u)\}^2 .$$

The score function $J(u) = \Phi^{-1}(u)$.

The exponential distribution with

$$f(x) = e^{-x}, \quad 1 - F(x) = e^{-x}$$

has quantile function

$$Q(u) = \log \frac{1}{1-u}$$

since $u = F(x)$, $x = Q(u)$ implies $1 - u = e^{-Q(u)}$; $fQ(u) = 1 - u$.

To illustrate how one obtains distribution functions from quantile functions, consider

$$Q(u) = \tfrac{1}{2}(\log \frac{1}{1-u})^{\frac{1}{2}} .$$

Consequently, $x = Q(u)$ implies $u = F(x) = 1 - e^{-4x^2}$, and $f(x) = 8x\, e^{-4x^2}$ which is the Rayleigh distribution; $fQ(u) = 4(1 - u)(-\log(1 - u))^{\frac{1}{2}}$.

An important consequence of the Correspondence and Inverse Identities are the following facts. Let U denote a random variable uniformly distributed on the interval [0, 1], and $\sim$ denote "identically distributed as." Then one can represent X by the

$$\text{Representation Identity:} \quad X \sim Q(U) . \tag{9}$$

When F is continuous

$$\text{Probability Integral Transformation:} \quad F(X) \sim U . \tag{10}$$

The representation identity plays a central role in theorem-proving because it enables one to first prove theorems for uniformly distributed random variables, and then extend to arbitrary random variables by using the representation (9) and the analytic properties of Q. This technique was first used by Scheffé and Tukey (1945). Tukey (1965) calls Q the representing function and q the sparsity function.

To simulate a sample $X_1, \ldots, X_n$ of a random variable X, one could simulate a sample $U_1, \ldots, U_n$ of U, and form $X_1 = Q(U_1), \ldots, X_n = Q(U_n)$. I understand from Professor Jim Thompson of Rice that the numerical analysis techniques are now available to make this a practical universal approach to simulating an arbitrary continuous distribution.

One may want to transform X to another random variable Y which has a specified continuous distribution function $F_0(x)$ and quantile function $Q_0(u)$. This can be accomplished using the facts

$$\text{Transformation Identities:} \quad X \sim QF_0(Y) \ , \quad Y \sim Q_0F(X) \ . \tag{17}$$

<u>Quantile Function of a Monotone Function</u>. An extremely useful property of the quantile function is how easily it can be found for a random variable Y which is obtained from X by a monotone transformation g: $Y = g(X)$. When g is an increasing function,

$$Q_Y(u) = g(Q_X(u)) \ . \tag{12}$$

When g is a decreasing function,

$$Q_Y(u) = g(Q_X(1-u)) \ . \tag{13}$$

Applications of these formulas are

$$Y = \mu + \sigma X \ , \qquad Q_Y(u) = \mu + \sigma Q_X(u) \ , \tag{14}$$

$$Y = -\log X \ , \qquad Q_Y(u) = -\log Q_X(1-u) \ , \tag{15}$$

$$Y = 1/X \ , \qquad Q_Y(u) = 1/Q_X(1-u) \ . \tag{16}$$

<u>Moments</u>: Moments of X are easily expressed in terms of the quantile function since

$$\text{Moment Identity: } E[g(X)] = E[g(Q(u))] = \int_0^1 g(Q(u))du \, . \tag{17}$$

The mean is

$$\mu = \int_0^1 Q(u)du \, . \tag{18}$$

The median is Q(0.5). Parameters whose estimation is considered more robust are the trimean $\frac{1}{4}Q(0.25) + \frac{1}{2}Q(0.5) + \frac{1}{4}Q(0.75)$, the trimmed mean

$$\mu_p = \int_p^{1-p} Q(u)du \, , \tag{19}$$

and the weighted mean $\mu_w = \int_0^1 w(u)Q(u)du$, for a specified weight function $w(u)$. Corresponding measures of variation would be functionals of the deviations $|Q(u) - \mu_w|$ of the quantile function from the representative value μ_w. The variance can be expressed

$$\sigma^2 = \int_0^1 \{Q(u) - \mu\}^2 du \, . \tag{20}$$

<u>Tail behavior of probability laws</u>. To describe all possible continuous probability distributions, it suffices to describe all possible continuous monotone functions $Q(u)$, $0 \le u \le 1$. To describe all possible types of tail behavior of probability laws, it suffices to describe the behavior of $Q(u)$ as u tends to 0 or 1. We choose to express this behavior in terms of $fQ(u)$. Let α be a parameter satisfying $-\infty < \alpha < \infty$. We call α

(i) lower tail exponent if

$$fQ(u) \sim u^{\alpha} \quad \text{as} \quad u \to 0 \, ,$$

$$\alpha = \lim_{u \to 0} \frac{uJ(u)}{fQ(u)}$$

(ii) upper tail exponent if

$$fQ(u) \sim (1 - u)^{\alpha} \quad \text{as} \quad u \to 1 ,$$

$$\alpha = \lim_{u \to 1} \frac{(1-u)J(u)}{fQ(u)} .$$

The parameter ranges (i) $0 \leq \alpha < 1$, (ii) $\alpha = 1$, and (iii) $\alpha > 1$ correspond to the three types of tail behavior

(i) short tails or limited type,

(ii) medium tails or exponential type,

(iii) long tails or Cauchy type.

The parameter range $\alpha < 0$ could be called super-short tails; the densities are unbounded and the corresponding characteristic functions $\psi(t) = E[e^{itx}]$, $-\infty < t < \infty$, decay very slowly as $t \to \infty$ and are not even integrable. The general treatment of such random variables require further study. (An example is $X = \cos \pi U$).

Extreme value distributions are those corresponding to the random variables,

$$\text{(i) } \xi^{\beta} , \quad \text{(ii) } \log \xi , \quad \text{(iii) } \xi^{\beta}$$

where ξ is exponential with mean 1, and β depends on the value of α, $\beta = 1 - \alpha$. The quantile functions of Weibull distributions are

$$Q(u) = (\log \frac{1}{1-u})^{\beta} , \quad \beta = 1 - \alpha \quad \text{where} \quad 0 \leq \alpha < 1 ;$$

the extreme value distribution has quantile function

$$Q(u) = \log \log \frac{1}{1-u} , \alpha = 1 .$$

Note that for $\beta = 1$, corresponding to $\alpha = 0$, the distribution is exponential.

<u>Conditional means and an approach to Empirical Bayes Estimation.</u> When the distribution of observations X depend on an unknown parameter θ to be estimated, a Bayesian estimator of θ is the conditional mean $E[\theta|X]$ usually written

$$E[\theta|X] = \frac{\int_{-\infty}^{\infty} \theta f(X|\theta)g(\theta)d\theta}{\int_{-\infty}^{\infty} f(X|\theta)g(\theta)d\theta}$$

where $f(X|\theta)$ is the conditional density of X given θ, and $g(\theta)$ is the prior density of θ. Let $Q_\theta(u)$ denote the prior quantile function of θ;

$$Q_\theta(u) = G^{-1}(u) \ , \quad G(x) = \int_{-\infty}^{x} g(\theta)d\theta \ .$$

One can show that

Conditional Mean Identity:
$$E[\theta|X] = \frac{\int_0^1 Q_\theta(u)f(X|Q_\theta(u))du}{\int_0^1 f(X|Q_\theta(u))du} \ .$$

In practice, one may be willing to assume $f(X|\theta)$ but the prior distribution of θ is unknown. The empirical Bayes attitude is to estimate the distribution of θ from previous estimators $\hat{\theta}_1, \ldots, \hat{\theta}_n$. What our new formula for $E[\theta|X]$ indicates is that it suffices to estimate the prior quantile function $Q_\theta(u)$.

The conditional distribution of θ given the observations X can be evaluated using the formula, for $0 \leq p \leq 1$

$$P[\theta \leq Q_\theta(p)|X] = \frac{\int_0^p f(X|Q_\theta(u))du}{\int_0^1 f(X|Q_\theta(u))du} \qquad (21)$$

To obtain a formula for the conditional quantile function of θ given X, denoted $Q_{\theta|X}(u)$, denote the right hand side of (21) by $D_{\theta|X}(p)$:

$$D_{\theta|X}(p) = \frac{\int_0^p f(X|Q_\theta(u))du}{\int_0^1 f(X|Q_\theta(u))du}, \quad 0 \leq p \leq 1 . \tag{22}$$

Then (21) can be written

$$F_{X|\theta}(Q_\theta(p)) = D_{\theta|X}(p) . \tag{23}$$

Then $F_{X|\theta}(x) = u$ for $x = Q_\theta(p)$ where p satisfies $D_{\theta|X}(p) = u$ whence $p = D^{-1}_{\theta|X}(u)$ and

Conditional Quantile Identity:
$$Q_{\theta|X}(u) = Q_\theta(D^{-1}_{\theta|X}(u)) \tag{24}$$

which is an extremely important formula.

We might regard the conditional median $Q_{\theta|X}(0.5)$ as an estimator of θ given X; (24) says that it equals the prior quantile function Q_θ evaluated at $D^{-1}_{\theta|X}(0.5)$.

One may be able to quickly obtain insight into whether a new observation X implies a "significantly different" estimator of θ. Speaking extremely intuitively, we can form an acceptance region for the null hypothesis H_0 that the "true" value of θ is approximately the prior median $Q_\theta(0.5)$ at a level of significance α; define $p = D^{-1}_{\theta|X}(0.5)$ and call it the p-median of the observation X. Accept H_0 if p satisfies an inequality of the form $\alpha/2 \leq p \leq 1 - (\alpha/2)$, $p \leq 1 - \alpha$, $p \geq \alpha$ (depending on whether the test is two-sided or one-sided).

Other consequences of "thinking quantile". A $(1 - \alpha)$-confidence level "prediction" interval for the values of a random variable X with a symmetric distribution could be the interval $Q(\alpha/2) \leq X \leq Q(1 - (\alpha/2))$. For a normal random variable $Q(u) = \mu + \sigma\Phi^{-1}(u)$ and the prediction interval is $|X - \mu| \leq \sigma\Phi^{-1}(\alpha/2)$. If X is an unbiased estimator $\hat{\theta}$ of a parameter θ, the confidence interval is $|\hat{\theta} - \theta| \leq \sigma\Phi^{-1}(\alpha/2)$ where σ is the standard deviation of $\hat{\theta}$. Many statistics text books use the intuitive notation $z(\alpha/2)$ for the mathematically precise $\Phi^{-1}(\alpha/2)$.

It should be noted that the shape of fQ and q is independent of location and scale parameters in the sense that the following equations are equivalent

$$F(x) = F_0(\frac{x-\mu}{\sigma}) \ , \quad Q(u) = \mu + \sigma Q_0(u) \ ,$$

$$f(x) = \frac{1}{\sigma} f_0(\frac{x-\mu}{\sigma}) \ , \quad fQ(u) = \frac{1}{\sigma} f_0Q_0(u) \ .$$

A symmetric density

$$f(x) = f(-x)$$

is equivalent to

$$Q(1 - u) = -Q(u) \ , \quad fQ(1 - u) = fQ(u) \ .$$

2. Sample Quantile Function

The basic definition of the sample quantile function $\tilde{Q}(u)$, $0 \leq u \leq 1$, is

$$\tilde{Q}(u) = \tilde{F}^{-1}(u) = \inf\{x: \tilde{F}(x) \geq u\}$$

where $\tilde{F}(x)$ is the sample distribution function. Its probability theory is explored in elegant detail in recent work of Czorgo and Revesz (1978) who show that

$$fQ(u)\{Q(u) - \tilde{Q}(u)\} \sim B(u) \ , \text{ a Brownian Bridge process .}$$

For data analysis we prefer a definition of $\tilde{Q}(u)$ which is differentiable so that a sample quantile-density function

$$\tilde{q}(u) = \tilde{Q}'(u)$$

can be defined. Its basic property is that

$$\tilde{q}(u) \sim \text{exponential with mean } q(u) .$$

In this section we emphasize the computational details involved in computing $\tilde{Q}(u)$ under various definitions. An alternative name and notation for the sample quantile function $\tilde{Q}(u)$, $0 \leq u \leq 1$ is the percentile function $X(p)$, $0 \leq p \leq 1$. It can be given a variety of slightly different definitions, which are all equivalent in the limit as the sample size tends to ∞. Intuitively, $X(p)$ is a value such that $100p\%$ of the numbers in the sample are less than or equal to $X(p)$, and $100(1 - p)\%$ of the numbers in the sample are greater than or equal to $X(p)$. Mathematically one might define $X(p)$ as the inverse of the sample distribution function, denoted by $\tilde{F}(x)$ or $P(x)$:

$$X(p) = P^{-1}(p) = \inf\{x: P(x) \geq p\} , \qquad 0 \leq p \leq 1 .$$

When P is continuous, $P(X(p)) = p$. When $P(x)$ is also strictly increasing, $x = X(p)$ if and only if $p = P(x)$.

When the sample of size n is available as a set of numbers (the case of a histogram is considered next), one can form the order statistics of the sample, denoted

$$X_{1;n} \leq X_{2;n} \leq \ldots \leq X_{n;n} ;$$

the order statistics are the numbers in the sample arranged in non-decreasing order. To define $X(p)$ we associate with p an index $i(p)$; possible definitions for $i(p)$ are

$$i(p) = np + \tfrac{1}{2} \quad \text{or} \quad i(p) = (n + 1)p \; .$$

We consider explicitly only the first definition of $i(p)$. When $i(p)$ is an integer, we define

$$X(p) = X_{i(p);n} \; .$$

This is equivalent to defining the value of $X(p)$ for $p = (2j - 1)/2n$ to be $X_{j;n}$. For other values of p we define $X(p)$ to be either piecewise constant or by linear interpolation. The piecewise constant definition is most convenient for hand calculation; it is equivalent to: if $i(p)$ is not an integer, choose an integer j so that $j < i(p) < j + 1$, and define $X(p) = \tfrac{1}{2}\{X_{j;n} + X_{(j+1);n}\}$. The linear interpolation definition is preferred in general because its derivative exists.

We argue that all the statistical information in a sample is contained in (and indeed is easily extracted from) the percentile function. Examples of how easily statistical measures are expressed in terms of percentile functions are:

median	$X(0.5)$
quartiles	$X(0.75$, $X(0.25)$
mid-quartile	$\tfrac{1}{2}\{X(0.25) + X(0.75)\}$
tri-mean	$\tfrac{1}{4}X(0.25 + \tfrac{1}{2}X(0.5) + \tfrac{1}{4}X(0.75)$
mean	$\bar{X} = \int_0^1 X(p)dp$

In words, the mean of the sample is the mean of the percentile function.

p-trimmed mean	$\frac{1}{1-2p}\int_{-p}^{p} X(u)du$
p-Winsorized mean	$p\ X(p) + \int_{p}^{1-p} X(u)du + p\ X(1-p)$
Interquartile range	$X(0.75) - X(0.25)$
Standard deviation	$SD = \sqrt{\text{Variance}}$
Variance	$\int_0^1 (X(p) - \bar{X})^2\, dp$

In words, the standard deviation is the root mean square of the deviations of the percentile function from the representative value $\bar{X}$.

Wilcoxon test statistic	$\int_0^1 p \text{ sign } X(p)\, dp$
Sign test statistic	$\int_0^1 \text{sign } X(p)\, dp$

<u>Histograms</u>. When a sample of size n is described by a histogram, one defines k intervals by endpoints x_{j-1}, x_j for $j = 1, \ldots, k$. The number of observations with values in $x_{j-1} < x \leq x_j$ is called the <u>frequency</u> of the interval. For $j = 1, \ldots, k$, define

$$p_j = \frac{n_j}{n}, \quad f_j = \frac{p_j}{x_j - x_{j-1}}$$

called respectively the <u>relative frequency</u> and the <u>density</u> in the j-th interval. To describe a sample, one gives the table:

Interval	Frequency	Relative Frequency	Cumulative Relative Frequency	Density
x_{j-1} to x_j	n_j	p_j	$p_1 + p_2 + \ldots + p_j$	f_j

and plots the histogram function h(x) defined by

$$h(x) = f_j\,, \qquad x_{j-1} < x < x_j$$
$$= \tfrac{1}{2}\{f_j + f_{j+1}\}\,, \qquad x = x_j\,.$$

The sample distribution function F(x) is defined to be integral of the histogram:

$$\tilde{F}(x) = \int_{-\infty}^{x} h(y)dy\,.$$

$\tilde{F}(x)$ is computed as follows: for $j = 1, 2, \ldots, k$

$$\tilde{F}(x_j) = p_1 + p_2 + \ldots + p_j\,.$$

At other values of x, define $\tilde{F}(x)$ by linear interpolation. Thus for $x_{j-1} \le x \le x_j$

$$\tilde{F}(x) = \tilde{F}(x_j)\,\frac{(x-x_{j-1})}{(x_j-x_{j-1})} + \tilde{F}(x_{j-1})\,\frac{(x_j-x)}{(x_j-x_{j-1})}\,.$$

Given an histogram table the percentile or sample quantile function is defined by

$$\tilde{Q}(0) = x_0\,,$$
$$\tilde{Q}(u) = x_j \quad \text{for} \quad u = p_1 + p_2 + \ldots + p_j$$

and by linear interpolation for other values of u in $0 < u < 1$.

It is to be emphasized that computing and plotting the sample quantile function is the first step in statistical data analysis, expecially when combined with the box-plot technique of Tukey (1977) to provide Quantile-Box Plots (Parzen (1979)).

3. Raw Density-Quantile Function $\tilde{f}Q$.

Having computed and plotted $\tilde{Q}$, the next step in statistical data analysis is to compute a raw estimator of the density-quantile function which we denote by $\tilde{f}Q(u)$.

When the data is reported as a histogram, the raw density quantile function, denoted $\tilde{f}Q(u)$, is defined by

$$\tilde{f}Q(u) = h(\tilde{Q}(u)) \;;$$

it satisfies for u in $p_1 + \ldots + p_{j-1} < u < p_1 + \ldots + p_j$

$$\tilde{f}Q(u) = f_j \;.$$

For $u = p_1 + \ldots + p_j$, $\tilde{f}Q(u) = \frac{1}{2}(f_j + f_{j+1})$.

When the data is reported as $\tilde{Q}$, a raw density-quantile function is formed from a raw estimator $q^*(u)$ of the quantile-density function as follows:

$$\tilde{f}Q(u) = 1/q^*(u) \;.$$

The basic requirement for $q^*(u)$ is that it be slightly smoother than $\tilde{q}(u)$. In general, to form a smooth estimator $q^*(u)$ of $q(u)$ one could use a kernel estimator

$$q^*(u) = \int_0^1 \tilde{q}(p)\, \frac{1}{h}\, K\!\left(\frac{u-p}{h}\right) dp$$

for a suitable kernel K and band width h.

At this stage we only seek to smooth $\tilde{q}$ enough so that it would be statistically meaningful to form its reciprocal. Therefore, we recommend computing q^* at equi-spaced values $u = h, 2h, \ldots, 1-2h, 1-h$ by

$$q^*(jh) = \{\tilde{Q}((j+1)h) - \tilde{Q}((j-1)h)\} \div 2h \;.$$

Define $q^*(u)$ for other values of u by linear interpolation.

The properties of $q^*(u)$ are given by Bofinger (1975) who shows that it is asymptotically normal with asymptotic variance and mean

$$\mathrm{Var}[q^*(u)] = \frac{1}{2nh} q^2(u)$$

$$E[q^*(u)] = q(u) + \frac{1}{6} h^2 q''(u)$$

$$\mathrm{Bias}[q^*(u)) = \frac{1}{6} h^2 q''(u) .$$

Let h_{min} denote the choice of h which minimizes mean square error,

Mean Square Error = Variance + Bias Squared ;

one can show that one should choose h_{min} so that

Variance = 4 Bias Squared

whence

$$\frac{1}{2nh_{min}} (q(u))^2 = \frac{1}{9} h^4_{min} (q''(u))^2 .$$

The following important conclusion has been proved:

$$h_{min} = (\frac{1}{n})^{1/5} C(u)$$

where

$$C(u) = (4.5)^{1/5} (\frac{q(u)}{q''(u)})^{2/5} .$$

We seek a lower bound for h_{min} to yield reasonably accurate estimators. One can argue that a "worst" case is the Cauchy distribution for which $fQ(u) = (\sin \pi u)^2/\pi$, and C(u) has values given by the following table (taken from Bofinger (1975)):

u	0.50	0.60	0.70	0.80	0.90	0.95
C(u)	.41	.37	.28	.19	.11	.06

For $n = 2^5 = 32$, $h_{min} = \frac{1}{2}C(u)$; for $n = 4^5 = 1024$, $h_{min} = \frac{1}{4}C(u)$.

What we would like to do in practice is to compute $\tilde{f}Q(u)$ at an equi-spaced grid of values h, 2h, 3h, ..., 1 = 2h, 1 - h. A choice of h = 0.05 or 0.025 yields the amount of smoothing that is reasonable for the worst case (long tailed densities). The optimal choice of h would undoubtedly be larger (especially for values of u near 0.5). The path we follow to obtain an optimally smoothed estimator is to use preflattened smoothing, defined in the next section.

4. Smoothed Density-Quantile Function $\hat{f}Q$.

One can develop many approaches to forming smooth functions $\hat{q}(u)$ and $\hat{f}Q(u)$ which can be regarded as estimators of q(u) and fQ(u). The approach we recommend has three important features: (1) it smooths a pre-flattened sample quantile-density, (2) it uses autoregressive smoothers, and (3) it provides goodness-of-fit tests for hypotheses that the true distribution function belongs to a specified location and scale parameter family of distribution functions.

Goodness of Fit Tests. Goodness of fit tests are concerned with testing hypotheses about the distribution function F(x) of a random variable X. Let $F_0(x)$ be a specified distribution function with quantile function $Q_0(u)$, and density- quantile function $f_0Q_0(u)$. The unrealistic case of a simple hypothesis

$$H_0: F(x) = F_0(x) , \quad Q(u) = Q_0(u)$$

is considered first to illustrate how one formulates goodness of fit tests. Conventional statistics tests recommend transforming X to $U = F_0(X)$, and testing whether U is uniform, using tests based on the sample distribution function of U.

Our first departure from the conventional approach is to emphasize using tests based on the sample quantile function of U, which we denote by $\tilde{D}(u)$. One can express it in terms of the sample quantile function $\tilde{Q}(u)$ of X by

$$\tilde{D}(u) = F_0(\tilde{Q}(u)) .$$

A more realistic hypothesis for F(x) is a location-scale parameter model,

$$H_0: F(x) = F_0(\frac{x-\mu}{\sigma}) , \quad Q(u) = \mu + \sigma Q_0(u) .$$

Let $\hat{\mu}$ and $\hat{\sigma}$ be estimators of the unknown parameters. Conventional texts recommend forming

$$\hat{U}_1 = F_0(\frac{X_1-\hat{\mu}}{\hat{\sigma}}), \ldots, \hat{U}_n = F_0(\frac{X_n-\hat{\mu}}{\hat{\sigma}}) ,$$

and using tests based on their sample distribution function. We would prefer tests based on the sample quantile function which we now denote $\tilde{D}_0(u)$; it can be expressed:

$$\tilde{D}_0(u) = F_0(\frac{\tilde{Q}(u)-\hat{\mu}}{\hat{\sigma}}) .$$

A method of generalizing this procedure, which avoids estimating μ and σ, is suggested by forming the density

$$\tilde{d}_0(u) = \tilde{D}_0'(u)$$
$$= f_0(\frac{\tilde{Q}(u)-\hat{\mu}}{\hat{\sigma}})\tilde{q}(u) \frac{1}{\hat{\sigma}}$$

where $\tilde{q}(u) = \tilde{Q}'(u)$ is the sample quantile-density function.

An important formula for $\tilde{q}(u)$ is: for $j = 1, 2, \ldots, n-1$

$$\tilde{q}(u) = n(X_{j;n} - X_{j-1;n}) , \quad \frac{2j-1}{2n} < u < \frac{2j+1}{2n} ;$$

the values of $\tilde{q}(u)$ are called spacings. It should be noted that the statistical properties of $\tilde{q}(u)$ are isomorphic to those of the sample spectral density of a stationary time series.

A new approach is to define a new density function

$$\tilde{d}(u) = f_0Q_0(u)\tilde{q}(u) \frac{1}{\tilde{\sigma}_0}$$

where

$$\tilde{\sigma}_0 = \int_0^1 f_0Q_0(u)\tilde{q}(u)du .$$

Note that if $f_0\tilde{Q}_0(u)Q(u) = 0$ at $u = 0$ and 1, one can write

$$\tilde{\sigma}_0 = \int_0^1 J_0(u)\tilde{Q}(u)du$$

which is a scale estimator that often coincides with the usual estimator when H_0 holds.

We call $\tilde{d}(u)$ the weighted spacings function,

$$\tilde{D}(u) = \int_0^u \tilde{d}(t)dt$$

the cumulative weighted spacings function, and

$$\tilde{\phi}(v) = \int_0^1 e^{2\pi iuv} \tilde{d}(u)du , \quad v = 0, \pm 1, \ldots$$

the pseudo-correlations.

We can regard $\tilde{d}(u)$ as an "estimator" (unfortunately, only consistent when used as the integrand of an integral) of

$$d(u) = f_0 Q_0(u) q(u) \frac{1}{\sigma_0}$$

where

$$\sigma_0 = \int_0^1 f_0 Q_0(u) q(u) du \ .$$

Under the null hypothesis, $d(u)$ is identically 1.

Parzen (1979) introduces autoregressive estimators $\hat{d}(u)$ of $d(u)$ which can estimate $\hat{d}(u) = 1$ a specified proportion of the time when H_0 is true. One thus simultaneously tests whether H_0 is true, and estimates $d(u)$ when H_0 is rejected.

<u>Autoregressive Estimation of the Density Quantile Function</u>. To obtain an estimator $\hat{fQ}(u)$ of $fQ(u)$ which has good mean square error properties at each point u, and is not too wiggly as a function of u, it is desirable to use a parametric representation of fQ(u) to estimate it. The hypothesis H_0: $Q(u) = \mu + \sigma Q_0(u)$ is equivalent to the representation

$$fQ(u) = \frac{1}{\sigma} f_0Q_0(u) \ .$$

A more general representation is

$$fQ(u) = C_m|1 + \alpha_m(1)e^{2\pi iu} + \ldots + \alpha_m(m)e^{2\pi ium}|^2 \ f_0Q_0(u)$$

for some integer m, $C_m > 0$, and complex coefficients $\alpha_m(1), \ldots, \alpha_m(m)$. The "base" function $f_0Q_0(u)$ can often be suggested by the data through an inspection of $\tilde{fQ}(u)$. One would like to choose $f_0Q_0(u)$ so as to reduce the number m of parameters in the representation. One can show that under rather general conditions to any specified f_0Q_0 there exists (in the limit as m tends to ∞) a representation for fQ of the foregoing form.

The foregoing representation for fQ implies that d(u) has the representation (for some $K_m > 0$)

$$d(u) = K_m|1 + \alpha_m(1)e^{2\pi iu} + \ldots + \alpha_m(m)e^{2\pi ium}|^{-2} \ .$$

In words, d(u) is the reciprocal of the square modulus of a polynomial. Such a representation may appear at first sight as unpromising. However it is equivalent to the Fourier transform

$$\phi(v) = \int_0^1 e^{2\pi iuv} \ d(u)du \ , \qquad v = 0, \pm 1, \ldots$$

satisfying a difference equation

$$\phi(v) + \alpha_m(1)\phi(1 - v) + \ldots + \alpha_m(m)\phi(m - v) = 0 , \qquad v > 0 ,$$

which can be used to determine the coefficients $\alpha_m(j)$ if $\phi(v)$ are known. Further, one can determine K_m by

$$K_m = \phi(0) + \alpha_m(1)\phi(1) + \ldots + \alpha_m(m)\phi(m) .$$

One can form a sequence $f\hat{Q}_m(u)$ of smooth estimators of $fQ(u)$ as follows. First form estimators $\tilde{\phi}(v)$ of $\phi(v)$. Second, for each m, determine coefficients $\hat{\alpha}_m(j)$, $j = 1, \ldots, m$, by solving the system of equations, with $v = 1, \ldots, m$,

$$\tilde{\phi}(v) + \hat{\alpha}_m(1)\tilde{\phi}(1 - v) + \ldots + \hat{\alpha}_m(m)\tilde{\phi}(m - v) = 0 .$$

Third, define

$$\hat{K}_m = \tilde{\phi}(0) + \hat{\alpha}_m(1)\tilde{\phi}(1) + \ldots + \hat{\alpha}_m(m)\tilde{\phi}(m)$$

$$\hat{d}_m(u) = K_m|1 + \hat{\alpha}_m(1)e^{2\pi iu} + \ldots + \hat{\alpha}_m(m)e^{2\pi ium}|^{-2}$$

Fourth, define

$$\hat{fQ}_m(u) = C_m|1 + \hat{\alpha}_m(1)e^{2\pi iu} + \ldots + \alpha_m(m)e^{2\pi imu}|^2 \; f_0Q_0(u)$$

where

$$C_m^{-1} = \int_0^1 |1 + \hat{\alpha}_m(1)e^{2\pi iu} + \ldots + \alpha_m(m)e^{2\pi ium}|^2 \; f_0Q_0(u)du .$$

The crucial question is the <u>order</u> <u>determination</u> problem; find a value of the order m, to be denoted $\hat{m}$, such that $\hat{d}_{\hat{m}}(u)$ is an "optimal" estimator of d(u) and $\hat{fQ}_m(u)$ is an "optimal" estimator of fQ(u). Further research needs to be done on this problem.

5. Parametric Smoothed Quantile Functions $\hat{Q}$.

Estimation of the fQ function only determines Q up to location and scale parameters. Thus the parametric model for a true quantile function

$$Q(u) = \mu + \sigma Q_0(u)$$

where Q_0 is known, and μ and σ are parameters to be estimated, can arise either from theory or as part of the process of fitting a smooth quantile function to a empirical quantile function $\tilde{Q}$.

Parzen (1979) discusses efficient estimation of the location and scale parameters μ and σ in the parametric model $Q(u) = \mu + \sigma Q_0(u)$ for the true quantile function Q. Equivalent to using a restricted set of order statistics $X_{np;n}, \ldots, X_{nq;n}$ (or a trimmed sample) is using the sample quantile function $\tilde{Q}(u)$, $p \leq u \leq q$. One can form asymptotically efficient estimators denoted $\mu_{p,q}$ and $\sigma_{p,q}$, using normal equations in suitable linear functionals in $\tilde{Q}$. Detailed formulas for these estimators, and their application to robust estimation, are discussed in Parzen (1979a).

References

Bofinger, E. (1975), "Estimation of a density function using order statistics," Austral. J. Statistics 17, 1-7.

Bofinger, E. (1975), "Non-parametric estimation of density for regularly varying distributions," Austral. J. Statistics 17, 192-195.

Czorgo, M. and Revesz, P. (1978), "Strong Approximations of the Quantile Process," Annals Statistics 6, 882-897.

Parzen, E. (1979), "Non-parametric Statistical Data Modeling," Journal American Statistical Association , 74, 105-131 (with discussion).

Parzen, E. (1979a), "A Density-quantile function perspective on robust estimation." Robustness in Statistics. ed. R. Launer and G. Wilkinson, New York: Academic Press.

Scheffé, H. and Tukey, J. W. (1945), "Non-parametric estimation, I Validation of order statistics," Ann. Math. Statist. 16, 187-192.

Tukey, J. N. (1965), "Which part of the sample contains the information," Proc. Nat. Acad. Sci. 53, 127-134.

GLOBAL MEASURES OF DEVIATION FOR KERNEL AND NEAREST NEIGHBOR DENSITY ESTIMATES

M. Rosenblatt*
University of California, San Diego
La Jolla, California 92032/USA

Abstract. A number of estimates of the probability density function (and regression function) have been introduced in the past few decades. The oldest are the kernel estimates and more recently nearest neighbor estimates have attracted attention. Most investigations have dealt with the local behavior of the estimates. There has, however, been some research and some heuristic comment on the utility of global measures of deviation like mean square deviation. Here, it is suggested that in a certain setting such global measures of deviation for kernel estimates may depend far less on tail behavior of the density function than in the case of nearest neighbor estimates. This appears to be due to the unstable behavior of the bias of nearest neighbor density estimates in the tails.

*This research is supported in part by ONR Contract N00014-75-C-0428.

Introduction. We first casually note some old local results for kernel estimates. *Let* $w(u)$ *be an integrable bounded weight function* with

$$\int w(u)du = 1 ,$$

and *let a sequence of bandwidths* $b(n) \downarrow 0$ as $n \to \infty$. Consider a sample $X_1,\ldots,X_n$ of *independent observations* from a population with *density function* $f(x)$. A one-dimensional kernel estimate $f_n(x)$ of $f(x)$ determined by w and $b(n)$ is given by

$$(1) \qquad f_n(x) = \frac{1}{nb(n)} \sum_{j=1}^{n} w\left(\frac{x-X_j}{b(n)}\right) .$$

If f *is continuous at* x *with* $f(x) > 0$ *the variance*

$$(2) \qquad \sigma^2[f_n(x)] \simeq \frac{f(x)}{nb(n)} \int w^2(u)du .$$

Also *if* f *and its first two derivatives are continuous and bounded* and $\int uw(u)du = 0$, $\int u^2 w(u)du < \infty$, then the bias

$$(3) \qquad E\, f_n(x)-f(x) = \frac{1}{2}\, b(n)^2\, f''(x) \int w(u)u^2du + o(b(n)^2) .$$

The local mean square error $E|f_n(x)-f(x)|^2$ under these conditions decreases to zero at a fastest rate $n^{-4/5}$ if the bandwidth $b(n) \simeq cn^{-1/5}$ as $n \to \infty$ [2].

A k-nearest neighbor density estimate is of the form

$$(4) \qquad f_n(x) = \frac{1}{nR_n} \sum_{j=1}^{n} w\left(\frac{x-X_j}{R_n}\right) ,$$

where $R_n = R_n(x)$ is the distance between x and the kth nearest neighbor to x among the X_j's. Also $k = k(n)$ *is a sequence of positive integers such that* $k \to \infty$ $k/n \to 0$ as $n \to \infty$. The following local estimates have been obtained for these estimates (see [1]). Assume that

$$\int |u|\; |w(u)|du < \infty ,$$

and that $P[z: |z-x| \geq r] = O(r^{-\alpha})$ for some $\alpha > 0$ as $r \to \infty$. Then *if* f *is bounded, continuously differentiable in a neighborhood of* x *with* $f(x) > 0$

$$(5) \qquad \sigma^2(f_n(x)) = \frac{2f^2(x)}{k} \int w^2(u)du + o\left(\frac{1}{k}\right)$$

as $n \to \infty$. Further, *if in addition*

$$\int u^2 |w(u)|du < \infty ,$$

and f *is continuously differentiable up to second order in a neighborhood of* x the bias

$$(6) \qquad E\, f_n(x)-f(x) = \frac{1}{8}\, \frac{f''(x)}{f(x)^2} \int u^2 w(u)\,du \left(\frac{k}{n}\right)^2 + \frac{2f(x)}{k}\, \frac{w(1)+w(-1)}{2} + o\left(\left(\frac{k}{n}\right)^2 + \frac{1}{k}\right)$$

as $n \to \infty$. The primary term in the bias in computing the mean square error at x is the first term on the right of (6). One again finds that the local mean square error decreases to zero at a fastest rate $n^{-4/5}$ if $k = k(n) \simeq cn^{4/5}$. The choice of an "optimum" c would require knowledge of $f(x)$, $f''(x)$ something which is a priori meaningless since the whole object is to estimate $f(x)$. Nonetheless, if one had this knowledge the asymptotic "optimum" behavior locally of both kernel and k-nearest neighbor with the same weight function would be the same. Of course one could use a bootstrap technique to approximate an "optimum" c by using crude initial estimates of $f(x)$, $f''(x)$. This could be adapted to a sequential procedure.

In comparing kernel and nearest neighbor estimates locally it is clear that k/n corresponds to $b(n)$. If $f(x)$ is large the variance of the kernel estimate is likely to be smaller than that of the k nearest neighbor estimate and the bias of the kernel estimate larger than that of the k nearest neighbor estimate. The relative magnitudes of variance and bias of the two types of estimates are reversed if $f(x)$ is small.

The remarks made above have been for the one dimensional estimates. However, corresponding estimates are available in the multidimensional context and analogous remarks can be made there.

A number of people (see [3] for regression estimates) have suggested using global measures of deviation like

$$(7) \qquad E\int |f_n(x)-f(x)|\,dx, \quad E\int |f_n(x)-f(x)|^2 dx ,$$

or weighted measures

$$(8) \qquad E\int |f_n(x)-f(x)|\,f(x)\,ds, \quad E\int |f_n(x)-f(x)|^2 f(x)\,dx .$$

We will be particularly interested in the behavior of these measures of deviation when the support of f is infinite. An important difference between kernel and nearest neighbor estimates can now be noted. In looking at global measures of deviation like (7) and (8), it's natural to ask whether estimates of local variance (see (2) and (5)) and of local bias (see (3) and (6)) are good approximations for fixed n over the whole infinite x domain. Under appropriate conditions this is true for kernel estimates but not for k-nearest neighbor estimates.

Global measures and kernel estimates. In [2] the global measure $\int E|f_n(x)-f(x)|^2 dx$ was examined for kernel estimates. The density f *was assumed to be bounded and twice continuously differentiable with* $f, f'' \in L^2$. Also w *was assumed to be non-negative* and *symmetric* with $\int w(u)u^2 du < \infty$. It was then indicated that if one took $b(n) = kn^{-1/5}$ with

$$k = \frac{2^{2/5}\left[\int w^2(v)dv\right]^{1/5}}{\left(\int (f''(x))^2 dx\right)^{1/5}\left(\int w(u)u^2 du\right)^{2/5}} \tag{9}$$

the following estimate for the integrated mean square error would hold

$$\int E|f_n(x)-f(x)|^2 dx$$

$$= 2^{3/5}\left[\int w^2(v)dv\right]^{4/5}\left[\int |f''(x)|^2 dx\left(\int w(v)v^2 dv\right)^2\right]^{1/5} n^{-4/5} \tag{10}$$

$$+ o(n^{-4/5})$$

as $n \to \infty$. The significant thing to notice here is that this global measure decreases to zero at the rate $n^{-4/5}$ as $n \to \infty$ under broad conditions on f and w. This is the same as the local rate at which the mean square error tends to zero for a fixed location x as $n \to \infty$. In particular, this rate of decrease for kernel estimates does not depend on a narrowly specified rate of decay for the density function estimated at infinity.

In a similar way we can analyze the behavior of $\int E|f_n(x)-f(x)|\, dx$ as $n \to \infty$ for kernel estimates. Now

$$E|f_n(x)-f(x)| = E|f_n(x)-E\, f_n(x)-(E\, f_n(x)-f(x))|$$

$$\le \sigma(f_n(x)) + |E(f_n(x)-f(x))| = \sigma(f_n(x)) + |b_n(x)| .$$

It is clear that

$$\sigma^2[f_n(x)] \le \frac{1}{nb(n)}\int w^2(v)f(x-b(n)v)dv \tag{11}$$

and

$$b_n(x) = \int_0^\infty w(v)\{f(x-b(n)v) + f(x+b(n)v) - 2f(x)\}dv . \tag{12}$$

Let us *assume that* w *is bounded, symmetric and bandlimited* and also that

$$f(x),\ |f'(x)|,\ |f''(x)| \le \frac{c}{1+|x|^{2+\varepsilon}}$$

for all x, where c and ε are positive constants. Then simple estimates using (11) and (12) indicate that

$$\int E\,|f_n(x)-f(x)|dx \le \frac{1}{\sqrt{nb(n)}}\left\{\int w^2(v)dv\right\}^{\frac{1}{2}}\int \sqrt{f(x)}\,dx$$

(13) $$+\ \frac{1}{2}\,b(n)^2\int w(v)v^2dv\int|f''(x)|dx$$

$$+\ o\left(\frac{1}{\sqrt{nb(n)}}+b(n)^2\right).$$

If $b(n) = kn^{-1/5}$ one finds that

(14) $$\int E|f_n(x)-f(x)|dx = O(n^{-2/5}).$$

The expression

(15) $$\int E|f_n(x)-f(x)|^2 f(x)dx = O(n^{-4/5}),$$

and

(16) $$\int E|f_n(x)-f(x)|f(x)dx = O(n^{-2/5}),$$

if $b(n) = kn^{-1/5}$ under the conditions used in discussing $\int E|f_n(x)-f(x)|^2dx$ and $\int E|f_n(x)-f(x)|dx$ respectively.

<u>A nearest neighbor estimate</u>. We shall only consider the simplest of kth nearest neighbor estimates in the one dimensional case (using heuristic arguments in part) though comparable results should hold for a large class of such estimates. The simplest k^{th} nearest neighbor estimate is

(17) $$f_n(x) = \frac{k-1}{n}\,\frac{1}{2R_n(x)}.$$

If we let

(18) $$u = G(t) = F(x+t)-F(x-t) = \int_{x-t}^{x+t} f(v)dv,$$

then

(19) $$E[f_n(x)]^s = \left(\frac{1}{2}\,\frac{k-1}{n}\right)^s\int \frac{1}{\{G^{-1}(u)\}^s}\,n\binom{n-1}{k-1}u^{k-1}(1-u)^{n-k}du.$$

Estimates of (19) for $s = 1$ or 2 will be required in guaging the size of some global measure of error. It is especially important to estimate $G^{-1}(u)$ or its reciprocal. Our estimates for $G^{-1}(u)$ or $1/G^{-1}(u)$ will depend on the relative magnitude of $1-F(x)$ and u. *Assume that* $(0,\infty)$ *is the support of* f.

The first alternative is the case

$$\text{(20)} \quad \text{(i)} \qquad 1-F(x) = o(u) \quad \text{as} \quad u \to o, \quad x \to \infty .$$

Then

$$u = F(x+t)-F(x-t) = 1 - F(x-t) - (1-F(x+t))$$

$$\sim 1 - F(x-t)$$

so that

$$x - t \simeq F^{-1}(1-u)$$

$$t \simeq x - F^{-1}(1-u) ,$$

and

$$\text{(21)} \qquad \frac{1}{t^s} = \frac{1}{\{G^{-1}(u)\}^s} \simeq \frac{1}{x^s}\left(\frac{1}{1-x^{-1}F^{-1}(1-u)}\right)^s .$$

Now $1 - F(x) = o(u)$ implies that $x^{-1}F^{-1}(1-u) = o(1)$ and so to the first orde

$$\text{(22)} \qquad \frac{1}{t^s} \simeq \frac{1}{x^s} .$$

In the second alternative

$$\text{(23)} \quad \text{(ii)} \qquad u = o(1-F(x)), \quad x \to \infty .$$

Assuming differentiability to the appropriate order we have

$$\text{(24)} \qquad u = G(t) = 2tf(x) + \frac{t^3}{3} f''(x) + \frac{2}{5!} t^5 f^{(4)}(x) + \ldots .$$

As a first order approximation one has

$$\text{(25)} \qquad t_{(1)} \simeq \frac{u}{2f(x)} .$$

Formally inserting $t = \frac{u}{2f(x)} + \alpha$ one obtains

$$o = 2f(x)\alpha + \frac{1}{3}\left(\frac{u}{2f(x)} + \alpha\right)^3 f''(x) + \ldots$$

so that to the appropriate order

$$\alpha = \frac{1}{6}\,\frac{1}{8}\,\frac{u^3}{f(x)^4} f''(x) + \ldots .$$

The second order approximation resulting is

$$(26) \qquad t_{(2)} = \frac{u}{2f(x)} - \frac{1}{48}\frac{f''(x)}{f(x)^4}u^3 .$$

Using this

$$\frac{1}{G^{-1}(u)} = \frac{1}{t} = \frac{1}{\frac{u}{2f(x)}}\left[1 + \frac{1}{24}\frac{f''(x)}{f(x)^3}u^2 + \dots\right]$$

$$= \frac{2f(x)}{u} + \frac{1}{12}\frac{f''(x)}{f(x)^2}u + \dots .$$

To get a third approximation let

$$t_{(3)} = t_{(2)} + \beta .$$

Then using (22) we find that

$$u - \frac{1}{24}\frac{f''(x)}{f(x)^3}u^3 + 2f(x)\beta + \dots$$

$$+ \frac{1}{3}f''(x)\left\{\frac{u^3}{8f(x)^3} - \frac{3u^2}{4f(x)^2}\frac{1}{48}\frac{f''(x)}{f(x)^4}u^3 + \dots\right\}$$

$$+ \frac{2}{5!}f^{(4)}(x)\left\{\frac{u^5}{32f(x)^5} + \dots\right\} + \dots = u .$$

This implies that

$$(27) \qquad \beta = \left\{\frac{1}{8\cdot 48}\frac{f''(x)^2}{f(x)^7} - \frac{1}{5!}\frac{1}{32}\frac{f^{(4)}(x)}{f(x)^6}\right\}u^5 .$$

A third alternative is

$$(28) \quad (iii) \qquad u \simeq (1-F(x)) \quad \text{as} \quad x \to \infty .$$

Under the assumption that $f(x)$ *is monotonic decreasing for sufficiently large* x one immediately has the following crude bounds

$$(29) \qquad \begin{aligned} &2tf(x-t) \ge G(t) \ge 2tf(x+t) \\ &\frac{2f(x-t)}{u} \ge \frac{1}{t} \ge \frac{2f(x+t)}{u} . \end{aligned}$$

We shall now consider approximating $E[f_n(x)]$ when

$$(30) \qquad 1 - F(x) = o\left(\frac{k}{n}\right) .$$

Partition the range of integration

(31) $$E[f_n(x)] = \int \frac{1}{2}\,\frac{k-1}{n}\,\frac{1}{G^{-1}(u)}\, n\binom{n-1}{k-1}\, u^{k-1}(1-u)^{n-k}du$$

$$= \int_{A(1-F(x))}^{1} + \int_{0}^{A(1-F(x))} = ① + ② .$$

Here A is a fixed large number. Then

(32) $$① \simeq \frac{k-1}{2n}\,\frac{1}{x}$$

since

$$\frac{1}{t} = \frac{1}{G^{-1}(u)} \simeq \frac{1}{x} \quad \text{if} \quad 1 - F(x) = o(u)$$

and this holds for $u \in [A(1-F(x)),1]$. In getting an estimate for ② notice that in the range $u \leq A(1-F(x))$, using (29) and with $f(y)$ monotone decreasing for large y we have

$$1/G^{-1}(u) = o\left(\frac{1}{u}\right) .$$

But then

$$② = o\left(\int_{0}^{A(1-F(x))} \frac{k-1}{n}\, n\binom{n-1}{k-1}\, u^{k-2}(1-u)^{n-k}du\right)$$

$$= o\left(\frac{k-1}{n}\,(1-F(x))\right) .$$

This implies that if $1 - F(y) = O\left(\frac{1}{y}\right)$ as $y \to \infty$ then

(33) $$E[f_n(x)] \sim \frac{k-1}{2n}\,\frac{1}{x}$$

if $1 - F(x) = o\left(\frac{k}{n}\right)$. A similar argument shows that

(34) $$E[f_n(x)^2] \simeq \left(\frac{k-1}{2n}\right)^2 \frac{1}{x^2}$$

if $1 - F(x) = o\left(\frac{k}{n}\right)$ and $1 - F(y) = O\left(\frac{1}{y^2}\right)$ as $y \to \infty$.

The case in which

(35) $$\frac{k}{n} = o(1-F(x))$$

can be dealt with similarly. Consider $E[f_n(x)]$ as given in the first line of (31). It can be written as

(36) $$E[f_n(x)] = \int_0^{\epsilon(1-F(x))} + \int_{\epsilon(1-F(x)}^1 \cdot = ① + ②$$

where ϵ is a small number. In the first integral we use the estimate

$$\frac{1}{t} = \frac{1}{G^{-1}(u)} = \frac{2f(x)}{u} + \frac{1}{12}\frac{f''(x)}{f(x)}u + \dots$$

since in the range $0 \leq u \leq \epsilon(1-F(x))$ u is small compared to $1 - F(x)$. Then

(37) $$① = f(x) + \frac{1}{24}\frac{f''(x)}{f(x)^2}\left(\frac{k}{n}\right)^2 + \dots .$$

On the other hand,

(38) $$|②| \leq \frac{k}{n}\exp\{-\delta n(1-F(x))\}$$

for some small $\delta > 0$. A similar estimate yields

(39) $$\sigma^2[f_n(x)] \simeq \frac{2f^2(x)}{k},$$

when (35) holds.

We first make a brief remark about

(40) $$E\int|f_n(x)-f(x)|dx$$

for the nearest neighbor density estimate (17). The estimate (21) implies that

(41) $$\sigma[f_n(x)] = o\left(\frac{1}{x}\right),$$

when $1 - F(x) = o\left(\frac{k}{n}\right)$. But then (41) and (33) imply that (40) diverges. Thus, one of the global measures that is well-defined for kernel estimates isn't finite for the nearest neighbor estimate (17). We shall discuss

(42) $$E\int|f_n(x)-f(x)|^2dx$$

in some small detail but not the other global measures with weight $f(x)$ since their analysis is quite similar to that of (42). In particular, the magnitude of (42) will be estimated for density functions $f(x)$ with support on $(0,\infty)$ such that

(43) $$f(x) \simeq x^{-\alpha-1}, \ \alpha > 0,$$

as $x \to \infty$. Let $A >> 1$ and $0 < \varepsilon << 1$. Estimates (33) and (41) imply that the contribution to (42) from

$$\int_{x>A(\frac{n}{k})^{1/\alpha}} E|f_n(x)-f(x)|^2 dx \sim \left(\left(\frac{k}{n}\right)^{2+\frac{1}{\alpha}}\right).$$

The estimates (37) and (39) imply that

$$\int_{x<\varepsilon(\frac{n}{k})^{1/\alpha}} E|f_n(x)-f(x)|^2 dx \sim \left(\left(\frac{k}{n}\right)^{2+\frac{1}{\alpha}} + \frac{1}{k}\right).$$

Under the assumption (43) one can show that

$$\frac{1}{t} \leq \frac{B}{x}$$

if $u \simeq 1 - F(x)$, $x \to \infty$, where B is a large constant. But this implies that the contribution to (42) from the integral over the range $\varepsilon(\frac{n}{k})^{1/\alpha}$ to $A(\frac{n}{k})^{1/\alpha}$ is $\sim\left(\left(\frac{k}{n}\right)^{2+}\right.$
Notice that the optimal rate of decay here appears to depend on α.

References

1. Y. P. Mack and M. Rosenblatt, "Multivariate k-nearest neighbor density estimates" to appear in the Journal of Multivariate Analysis.

2. M. Rosenblatt, "Curve estimates," Ann. Math. Stat., 1971, vol. 42, 1815-1842.

3. C. J. Stone, "Consistent nonparametric regression," Ann. Stat., 1977, vol. 5, 595-602.

SOME COMMENTS ON THE ASYMPTOTIC BEHAVIOR OF ROBUST SMOOTHERS

Werner Stuetzle and Yashaswini Mittal

Department of Statistics
Stanford University
Stanford, California 94305/USA

Abstract. In curve estimation, running M-estimates are a natural generalization of Kernel-type smoothers (moving averages). We find the rate of convergence that can be expected from these estimates and the leading bias and variance terms. We also explain the effect of twicing for Kernel-type smoothers and give some rationale for its use in robust curve estimation.

I. Introduction

In recent years there has been some interest in the development and evaluation of robust smoothers, smoothers that are insensitive to occasional gross errors in the data. (Tukey [1], Velleman [2], Mallows [3], Huber [4].) A straightforward way to obtain such smoothers is to run a robust location estimate over the data in the same way as one runs a mean over the data in the case of a (nonrobust) moving average smoother. One aim of this paper is to provide a little insight into the asymptotic behavior of a family of smoothers of this type, namely the running M-estimates.

An interesting procedure which has been suggested in connection with robust smoothers is "twicing": Smooth the series of observations, smooth the residuals, and use the sum of the two smooths as the final result. In the second part of the paper we explain the effect of twicing for Kernel-type smoothers and give some rationale for its use in robust curve estimation.

II. Asymptotics for Running M-Estimates

II.A. Set-up of the problem.

Given an unknown, suitably regular function f. We want to estimate $f(x_0)$ from observations $(x_1, y_1), \ldots, (x_n, y_n)$ where

(i) x_i i.i.d. uniformly distributed on $[x_0-\Delta, x_0+\Delta]$

(ii) $y_i = f(x_i) + \varepsilon_i$ ε_i i.i.d. with density ϕ, symmetric around 0 and suitably regular. (1)

This is a slightly unusual set-up. Usually one assumes to be given N observations with abscissas equispaced in $[-1, 1]$. Then the number $n (= N\Delta)$ of x_i are fixed and equispaced in $[x_0-\Delta, x_0+\Delta]$. It will be clear in the following section that assumption 1(i) simplifies the problem considerably. It also looks eminently reasonable that the results of section II.B should be unchanged if the x_i are taken fixed and equispaced.

Without restriction of generality we can assume $x_0 = 0$ and $f(x_0) = 0$. The estimate $\hat{\theta}_n$ of $f(0)$ is given implicitly by the following equation

$$\sum_{i=1}^{n} \psi(y_i - \hat{\theta}_n) = 0 \tag{2}$$

where $\psi(x)$ is antisymmetric around 0, monotone increasing and bounded. We assume that ψ is appropriately scaled and thus the problem of estimating the scale of the ε_i does not arise. (See Huber [5].)

As a criterion for the precision of our estimate we use the expected squared error

$$ESE = E(\hat{\theta}_n - f(x_0))^2 = E(\hat{\theta}_n^2) . \tag{3}$$

The right hand side of (3) tends to 0 if both $\Delta \to 0$ and $n \to \infty$. In the traditional case n and Δ are related by $n = N\Delta$. The optimal choice of Δ as a function of N then is such that ESE goes to 0 at the fastest possible rate. Even though for our set-up Δ and n can be chosen in any manner, we assume $n = N\Delta$ to make the results comparable.

II.B. Bias, Variance and Convergence Rate of the Estimate.

Our observations y_i are i.i.d. with density

$$\tau_\Delta(y) = \frac{1}{2\Delta} \int_{-\Delta}^{\Delta} \phi(y - f(x))dx . \tag{4}$$

We now expand the M-estimate in the familiar way (see, for example, Huber [5]).

$$\hat{\theta}_n = \frac{1}{nE_\phi \psi'} \Sigma\psi(y_i) + r_{n,\Delta} = D_1 + r_{n,\Delta} , \text{ say} . \tag{5}$$

The expected squared error thus becomes

$$ESE = E(\hat{\theta}_n^2) = E(D_1^2) + 2E(D_1 r_{n,\Delta}) + E(r_{n,\Delta}^2) . \tag{6}$$

For the moment we look only at the first term

$$E(D_1^2) = \frac{1}{n(E_\phi\psi')^2} E_\tau\psi^2 + \frac{(1 - \frac{1}{n})}{(E_\phi\psi')^2} (E_\tau\psi)^2 = V+B^2 , \text{ say} . \tag{7}$$

V can be interpreted as a variance term, B as a bias term.

We first look at the variance term. Expanding $\phi(y - f(x))$ in a Taylor series around $x=0$ and observing that several terms vanish, we get

$$\begin{aligned} E_\tau\psi^2 &= \frac{1}{2\Delta} \int \psi^2(y)\ \phi(y - f(x))dx\ dy \\ &= E_\phi\psi^2 + \frac{\Delta^2}{6} f'^2(0) \int \psi^2(y)\ \phi''(y)dy + o(\Delta^2) . \end{aligned} \tag{8}$$

Similarly, integration by parts gives us

$$E_\tau\psi = -\frac{1}{6}\Delta^2 f''(0)\int \psi(y)\,\phi'(y)dy + o(\Delta^2)$$

$$= \frac{1}{6}\Delta^2 f''(0)\, E_\phi\psi' + o(\Delta^2)\,. \tag{9}$$

Putting things together, we get

$$E(D_1^2) = \frac{1}{n}\frac{E_\phi\psi^2}{(E_\phi\psi')^2} + (1-\frac{1}{n})\quad \frac{1}{36}\Delta^4 f''^2(0)$$

$$+\frac{\Delta^2}{6n(E_\phi\psi')^2} f'^2(0)\int \psi^2(y)\,\phi''(y)dy$$

$$+ o(\Delta^4) + o(\frac{\Delta^2}{n})\,. \tag{10}$$

Note that the ESE can go to 0 no faster than $\frac{1}{n}$. This rate is achieved if we choose $\Delta \sim n^{-1/4}$, because then $E(D_1^2) \sim 1/n$ and, under suitable regularity conditions on ϕ, ψ, $r_{n,\Delta} = o(\frac{1}{n})$. The middle term in (6) also is $o(\frac{1}{n})$ due to Schwarz's inequality.

In the traditional situation n and Δ are related by $n = N\Delta$. So we would have to choose $\Delta \sim N^{-1/5}$, which would give $ESE \sim N^{-4/5}$.

So we can summarize our results as follows. If we choose $\Delta \sim n^{-1/4}$, we achieve the optimal rate $ESE \sim \frac{1}{n}$. For this choice of Δ, the leading bias term in the ESE is $\frac{1}{36}\Delta^4 f''^2(0)$, the leading variance term is $\frac{1}{n}\frac{E_\phi\psi^2}{(E_\phi\psi')^2}$. For the traditional situation, ($n = N\Delta$) the choice of $\Delta \sim N^{-1/5}$ gives $ESE \sim N^{-4/5}$.

In retrospect, these results are not surprising. The variance is just the usual asymptotic variance of an M-estimator (Huber, [5]). The bias and the rate of convergence are the same as one would obtain when using a Kernel smoother with rectangular Kernel. This makes sense, because in the M-estimator all observations have the same weight.

III. The Asymptotics of Twicing

Given a curve estimation procedure T, which operates on (x_i, y_i), i=1, ..., N, and produces a smooth $(x_i, \hat{y}_i)$, i=1, ..., N. We obtain "T twice" by the following steps.

(i) Compute the residuals $r_i = y_i - \hat{y}_i$.

(ii) Apply T to (x_i, r_i), i=1, ..., N and obtain corrections c_i.

(iii) Define $\hat{y}_i^c = \hat{y}_i + c_i$. Use $(x_i, \hat{y}_i^c)$, i=1, ..., N as the final result.

We will now discuss the asymptotic effect of twicing for the case where T is a Kernel-type smoother. We assume x_i, $i=1, \ldots, N$ to be equispaced in $[0, 1]$. Ignoring boundary effects, the vector $\hat{\underline{y}}(N, \lambda)$ of smoothed ordinates is then given by

$$\hat{\underline{y}}(N, \lambda) = \underline{y} * \underline{w}(N, \lambda) , \tag{11}$$

where $*$ denotes the convolution and the weight vector $\underline{w}(N, \lambda)$ is obtained by discretizing the Kernel $K_\lambda(x) = \lambda K(\lambda x)$. The degree of smoothing is determined by the choice of λ.

If $y_i = f(x_i) + \varepsilon_i$ and ε_i is distributed symmetrically around 0, then the bias of such a Kernel smoother at x_0 is (formally)

$$b_\lambda(x_0) = \Sigma \frac{1}{\lambda^K K!} f^{(K)}(x_0)\, m_K , \tag{12}$$

where m_K, the K-th moment of the Kernel, is defined by $m_K = \int K(t)\, t^K\, dt$. If $m_0 = 1$ and $m_1, \ldots, m_{\ell-1}$ vanish, then polynomials of order $\ell-1$ and less are estimated bias-free (up to discretization and boundary effects), and for ℓ times continuously differentiable f the leading bias term is $f^{(\ell)} m_\ell/(\lambda^\ell \ell!)$.

Twicing the smoother yields

$$\begin{aligned}\hat{y}^c(N,\lambda) &= \underline{w}(N,\lambda) * (\underline{y}-\underline{y} * \underline{w}(N,\lambda)) + \underline{y} * \underline{w}(N,\lambda)\\ &= \underline{y} * (2\underline{w}(N,\lambda) - \underline{w}(N,\lambda) * \underline{w}(N,\lambda)) .\end{aligned} \tag{13}$$

For Kernel-type smoothers twicing thus amounts to using the Kernel $K^* = 2K - K * K$ instead of K. Let $m_0, m_1, \ldots$ denote the moments of K and $m_0^*, m_1^*, \ldots$ denote the moments of K^*. We then have the following.

__Lemma__. If $m_0 = 1, m_1, \ldots, m_{\ell-1} = 0, m_\ell, \ldots, m_{2\ell} < \infty$, then $m_0^* = 1$, $m_1, \ldots, m_{2\ell-1} = 0, m_{2\ell} < \infty$.

__Proof__. By definition, $m_{2\ell}^* = \int K^*(t)\, t^{2\ell} dt = 2m_{2\ell} - \iint K(t)\, K(x-t)\, x^{2\ell} dt\, dx$ $= 2m_{2\ell} - \iint K(t)\, K(x)\, (x+t)^{2\ell}\, dt\, dx = 2m_{2\ell} - \sum_{K=0}^{2\ell} \binom{2\ell}{K} \iint K(t)\, K(x)\, x^K\, t^{2\ell-K}\, dt\, dx$.
Because of $m_1, \ldots, m_{\ell-1} = 0$ and the symmetry in (x,t) this yields

$$\begin{aligned} m_{2\ell}^* &= 2m_{2\ell} - \binom{2\ell}{\ell} \int K(t)\, K(x)\, x^\ell t^\ell dt - 2m_{2\ell} \\ &= - \binom{2\ell}{\ell} m_\ell^2 . \end{aligned}$$

In the same way we get $m_1^*, \ldots, m_{2\ell-1}^* = 0$.

This shows that for the case of T being a Kernel-type smoother that reproduces polynomials up to degree $\ell-1$, T twice reproduces polynomials up to degree

$2\ell-1$. In this case however, there is no motivation for twicing. One can use a Kernel with more vanishing moments from the beginning. This is even more true as twicing an optimal Kernel (optimal in the sense of Epanechnikov [6]) does not give an optimal Kernel of higher order.

The situation becomes different if one looks at robust smoothers. The running M-estimates show qualitatively the same asymptotic behavior as Kernel-type smoothers with a Kernel that has vanishing first and existing second moment. They reproduce straight lines, and, for suitably chosen window width, their expected squared error converges to 0 at a rate of $N^{-4/5}$. We expect the same to be true for all running location estimates. There are not many robust smoothers known that reproduce quadratic polynomials and thus can achieve a convergence rate of $N^{-8/9}$. One possibility are robust splines (Huber [4]). Another possibility would be local robust fitting of quadratic polynomials. Both procedures are computationally not very attractive. In view of our findings about the effect of twicing on Kernel-type smoothers, we conjecture that twicing of running robust location estimates gives the same effect much cheaper.

References

[1] Tukey, J. W., EDA Exploratory Data Analysis. Addison-Wesley (1977).

[2] Velleman, P. Robust nonlinear data smoothers: Definitions and recommendations. Natl. Acad. Sci. USA (1977), Vol. 74, No. 2, pp. 434-436.

[3] Mallows, C. Some theory of non-linear smoothers. To appear in Ann. Statist. (July 1980).

[4] Huber, P. J. Robust Smoothing. To appear in Proc. of the ARO Workshop on Robust Statistics (April 1978). Academic Press (in press).

[5] Huber, P. J., Robust estimation of a location parameter. Ann. Math. Statist. (1964), Vol. 35, No. 1, pp. 73-101.

[6] Epanechnikov, V. A., Nonparametric estimates of a multivariate probability density. Theor. Prob. Appl. (1969), Vol. 14, pp. 153-158.

CROSS-VALIDATION TECHNIQUES FOR SMOOTHING SPLINE FUNCTIONS IN ONE OR TWO DIMENSIONS

by

F. UTRERAS D.*

LABORATOIRE DE MATHEMATIQUES APPLIQUEES
UNIVERSITE SCIENTIFIQUE ET MEDICALE DE GRENOBLE
BOITE POSTALE 53X
38041 GRENOBLE CEDEX

Abstract

This paper aims to present some results on the asymptotic behaviour of matrix associated with certain types of spline functions and shows how these resu can be used to obtain a fast algorithm for choosing the smoothing parameter in th smoothing of noisy data by splines.

First, we give a general theorem on the behaviour of the eigenvalues associated with a spline function. The spline functions considered here are those defined by a variational formulation (for general theorems on these splines see [1] ,[11]).

Second, we use the preceding results to obtain an O(n) - algorithm to calculate the optimal smoothing polynomial spline using the generalized Cross-Validation Technique, in the case of equally spaced data. We obtain an $O(n^2)$ algorithm to perform this calculation in the case of non equally spaced data. Som very good numerical results are presented and a table of run times is given. We then introduce a method which calculates the solution (smoothed data) by a piecew smoothing and correction technique. This method allows us to treat a very importa amount of data.

We show too, how calculations can be made with two dimensional data (wi arbitrary data points) and generalize the piecewise calculations to this case. We also present a set of numerical results and a table of run times.

---

* New address : Departamento de Matemáticas
Universidad de Chile
Casilla 2777
Santiago
CHILE

1 - INTRODUCTION

Let us consider the problem of approximating the unknown function $f : \Theta \subset R^k \to R$ $k = 1,2$ supposed to be "smooth", if we know the values $f(t_i)$ $i = 1,2,\ldots,n$ measured with error at n different points of Θ, more precisely, let

(1.1) $$z_i = f(t_i) + \varepsilon_i \qquad i = 1,2,\ldots,n$$

where ε_i $i = 1,2,\ldots,n$ are random numbers satisfying :

$$E[\varepsilon_i] = 0 \qquad i = 1,2,\ldots,n$$

(1.2) $$E[\varepsilon_i \, \varepsilon_j] = 0 \qquad i \neq j$$

$$E[\varepsilon_i^2] = v_i^2 \qquad i = 1,2,\ldots,n$$

and $f \in X$, where X is a semi-hilbertian subspace of R^{Θ} with the associated semi-norm $((.,.))^{1/2}$ and Null space N.

Doing convenient hypothesis on $\{t_1,\ldots,t_n\}$ and N (see next section), our estimate of f is the smoothing spline $\sigma_{n,\tau}$ defined by :

(1.3) $$\tau((\sigma_{n,\tau},\sigma_{n,\tau})) + \frac{1}{n}\sum_{i=1}^{n}\alpha_i^2\,(z_i - \sigma_{n,\tau}(t_i))^2 = \min_{g \in X}\{\tau((g,g)) + \frac{1}{n}\sum_{i=1}^{n}\alpha_i^2\,(z_i - g(t_i))^2\}$$

where the weights α_i^2 $i = 1,2,\ldots,n$ are defined by :

(1.4) $$\alpha_i^2 = \frac{\dfrac{1}{v_i^2}}{\dfrac{1}{n}\displaystyle\sum_{j=1}^{n}\frac{1}{v_j^2}}$$

It is important to see that the α'_is are completely determined if we know the v'_is up to a multiplicative constant. It is also easy to see that $v_i = v$ $i = 1,2,\ldots,n$ implies that $\alpha_i = 1$ $i = 1,2,\ldots,n$

It is well known (see [6], [7]) that there is a unique solution to the problem (1.3) (under the hypotheses we will see in the next paragraph).

Now, let us call $y_i^{\tau} = \sigma_{n,\tau}(t_i)$ $i = 1,2,\ldots,n$. We also know that the application :

(1.5) $$z \to y^{\tau}$$

is a linear one-to-one transformation from R^n into itself, for each value of $\tau > 0$. Let us call $A_n(\tau)$ the associated matrix.

The main subject of this paper is the practical choice of τ, the tradeoff between the "roughness" of the solution, measured by $((g,g))$, and the infidelity to the data, measured by $\frac{1}{n}\sum_{i=1}^{n}\alpha_i^2\,(z_i - g(t_i))^2$. Wahba and others have proposed to choose τ as the minimizer of the (Generalized Cross-Validation) function :

(1.6)

$$V_n(\tau)=\frac{\frac{1}{n}\sum_{i=1}^{n}\alpha_i^2\,(z_i-\sigma_{n,\tau}(t_i))^2}{(1-\frac{1}{n}\operatorname{Tr}(A_n(\tau)))^2}$$

For theoretical properties of this estimator see [5], [6], [21], [24], [23], [25].

As it has be shown, the minimizer of (1.6) provides a very good estimation of the optimal parameter, even for relatively small samples, but the cost of performing this calculation was very high.

The algorithms that we develop here, allow us to treat very great samples at a very reduced cost. So, the algorithm we develop for cubic or quintic splines costs 1/n times the cost of classical algorithms ([5], [26]) in the case of equally spaced data.

Finally, we use this estimate to smooth noisy bidimensional data, with arbitrary data points.

2 - MATHEMATICAL BACKGROUND

Let R^{Θ} be the vector space of all real valued functions with domain $\Theta \subset R^k$ $(k = 1,2)$

A vector subspace X of R^{Θ} will be said to be semi-hilbertian, if there is a semi-inner-product defined over X with the associated semi-norm $((.,.))^{1/2}$ and Null space N such that X/N is a Hilbert space (with the induced topology).

Let us consider the topology of pointwise convergence over R^{Θ} and denote $R^{(\Theta)}$ as its topological dual. It is well known that the elements of $R^{(\Theta)}$ are the real valued functions defined over Θ having a finite support. Between R^{Θ} and $R^{(\Theta)}$ we define the product of duality :

(2.1) $$\langle \nu, f\rangle = \sum_{t\in\Theta} \nu(t)\, f(t) \qquad \nu\in R^{(\Theta)} \qquad f\in R^{\Theta}$$

The set of functions $\{\delta_a\}$ $a\in\Theta$ such that :

(2.2) $$\delta_a(t) = \begin{cases} 1 & t = a \\ 0 & t \neq a \end{cases}$$

is an algebraic basis of $R^{(\Theta)}$.

Let N° denote the "orthogonal" space of N in $R^{(\Theta)}$, defined by :

(2.3) $$N^\circ = \{\nu\in R^{(\Theta)} \quad \langle \nu, p\rangle = 0 \;, \text{ for all } p\in N\}$$

<u>Définition</u> : 1) A semi-hilbertian space X of R^{Θ} is said to be a semi-hilbertian subspace of R^{Θ} if :

1.a) the canonical injection of X/N into R^{Θ}/N is continuous.

1.b) N is finite dimensional

2) We say that X separates the points of Θ if for all set of points $\{t_1, \ldots, t_n\}$ of Θ and real numbers $z_1, z_2, \ldots, z_n$; there exists a function $v \in X$ such that :

(2.4) $$v(t_i) = z_i \quad i = 1,2,\ldots,n$$

all these notations allow us to establish the following result :

<u>Theorem 2.1</u> Let X be a semi-hilbertian subspace of R^Θ which sepparates the points of Θ and let $\{t_1,\ldots,t_n\}$ be n different points of Θ such that :

$$\{ v \in X \ / \ v(t_i) = 0 \quad i = 1,2,\ldots,n \} \quad \cap N = \{0\}$$

Finally, let $z_1, z_2, \ldots, z_n$ be real numbers and $\tau > 0$ a real parameter. Then we have the following result :

(a) There exists a unique solution to the problem :

(2.4) Minimize $((g,g))$
$g \in X$
$g(t_i) = z_i \quad i = 1,2,\ldots,n$

we shall call this solution the interpolating spline function σ_n.

(b) there exists a unique solution $\sigma_{n,\tau}$ to the problem

(2.5) $$\text{Minimize}_{g \in X} \ \{\tau((g,g)) + \frac{1}{n} \sum_{i=1}^{n} \alpha_i^2 (g(t_i) - z_i)^2\}$$

This element will be called the smoothing spline of parameter τ.

<u>Proof</u> : see Duchon [6], [7]

Let us denote by S_n the linear subspace of X defined by :

(2.6) $$S_n = \{\sigma \in X \ / \ \exists d_1, d_2, \ldots, d_n \in R \text{ such that } ((\sigma,v)) = \sum_{i=1}^{n} d_i \, v(t_i) \text{ for all } v \in X\}$$

The following facts can be easily shown (cf [1]).

(i) S_n is an n - dimensional linear space

(ii) $\sigma_n \in S_n$ and $\sigma_{n,\tau} \in S_n$, for all $\tau > 0$

Now we state a characterization theorem that will be useful later. For that purpose, we need the following definition :

<u>Definition</u> : A real valued function $K(.,.)$ defined over $\Theta \times \Theta$ is called a <u>Reproducing Kernel</u> of X, if :

(i) for all $\nu \in N^0$, the real valued function :

$$s \rightarrow \sum_{t \in \Theta} \nu(t) \, K(s,t)$$

belongs to X

(ii) for all $\nu \in N^0$, $v \in X$ we have the 'reproducing property'

$$((\sum_{t \in \Theta} \nu(t) \, K(.,t), v)) = \sum_{t \in \Theta} \nu(t) \, v(t)$$

Duchon [6] has shown that every semi-hilbertian subspace of R^{Θ} has at least one Reproducing Kernel.

Theorem 2.2 Under the hypothesis of theorem (2.1), if $K(.,.)$ is a reproducing kernel of X :

1) There exist real numbers $d_1, d_2, \ldots, d_n$; $\beta_1, \beta_2, \ldots, \beta_m$ such that :

$$\sigma_n(t) = \sum_{i=1}^{n} d_i K(t,t_i) + \sum_{k=1}^{m} \beta_k p_k(t)$$

and the d_i, β_k satisfy the following linear system :

$$\sum_{i=1}^{n} d_i K(t_j,t_i) + \sum_{k=1}^{m} \beta_k p_k(t_j) = z_j \qquad j = 1,2,\ldots,n$$

$$\sum_{k=1}^{m} p_\ell(t_k)\, d_k = 0 \qquad \ell = 1,2,\ldots,m$$

where $p_1, p_2, \ldots, p_m$ is a basis of N

2) There exist real numbers $d_1^\tau, d_2^\tau, \ldots, d_n^\tau$; $\beta_1^\tau, \beta_2^\tau, \ldots, \beta_m^\tau$ such that :

$$\sigma_{n,\tau}(t) = \sum_{i=1}^{n} d_i^\tau K(t,t_i) + \sum_{k=1}^{m} \beta_k^\tau p_k(t)$$

and the d_i^τ, β_k^τ satisfy the following linear system :

$$\frac{n\tau}{\alpha_i^2} d_i^\tau + \sum_{j=1}^{n} K(t_j,t_i)\, d_j^\tau + \sum_{k=1}^{m} p_k(t_i)\beta_k^\tau = z_i \qquad i=1,\ldots,n$$

$$\sum_{j=1}^{n} p_k(t_j)\, d_j^\tau = 0 \qquad k = 1,\ldots,m$$

Proof : See Duchon [7]

Now we get an explicit expression for $A_n(\tau)$.

Let $\sigma_n^1, \sigma_n^2, \ldots, \sigma_n^n$ be the canonical basis of S_n, defined by :

(2.7) $\sigma_n^i(t_j) = \begin{cases} 1 & i=j \\ 0 & i\neq j \end{cases}$

Let s_n, σ_n be two elements of S_n such that :

$$\sigma_n(t_i) = x_i \qquad\qquad i = 1,2,\ldots,n$$
$$s_n(t_i) = y_i$$

Then we have :

$$\sigma_n = \sum_{i=1}^{n} x_i \sigma_n^i$$

$$s_n = \sum_{i=1}^{n} y_i \sigma_n^i$$

and

(2.8) $$((\sigma_n, s_n)) = \sum_{i=1}^{n} \sum_{j=1}^{n} x_i y_j ((\sigma_n^i, \sigma_n^j))$$

if we set :

(2.9) $$\omega_{ij}^n = ((\sigma_n^i, \sigma_n^j))$$

and call Ω_n the matrix of elements ω_{ij}^n, we have :

(2.10) $$((\sigma_n, s_n)) = x^T \Omega_n y = y^T \Omega_n x$$

It is clear that Ω_n is a symmetrical semi-positive definite n x n matrix, having an m dimensional null space and so, at least m zero eigenvalues. Let us call $y_i^\tau = \sigma_{n,\tau}(t_i)$ i = 1,2,...,n. We then have :

(2.11) $$\tau\Omega_n y^\tau + \frac{1}{n} D_n y^\tau = \frac{1}{n} D_n z$$

where D_n is a diagonal matrix whose diagonal elements are the α_i's.

From (2.11) we easily get :

(2.12) $$A_n(\tau) = (I + n\tau D_n^{-1} \Omega_n)^{-1}$$

Let $\lambda_{1n}, \lambda_{2n}, \ldots, \lambda_{nn}$ be the n eigenvalues of $nD_n^{-1} \Omega_n$ in increasing order. Then ,

(2.13) $$\mathrm{Tr}\,(A_n(\tau)) = \sum_{i=1}^{n} \frac{1}{1 + \tau\lambda_{in}}$$

In the following paragraph we study the behaviour of the λ_{in}'s and get an asymptotic expression of (2.13).

3 - THE ASYMPTOTIC BEHAVIOUR OF THE EIGENVALUES

To obtain the results of this paragraph, we need the following hypotheses :

(3.1) $X \subset L^2(\Theta)$

(3.2) Θ is a bounded open set of R^k (k = 1,2) with regular boundary.

(3.3) X is a Hilbert space for the inner product :

$$((u,v)) + (u,v) \qquad u,v \in X$$

where (.,.) denotes the $L^2(\Theta)$ inner product.

Consider now the two following bilinear forms defined on X x X by :

(3.4) $$\mathcal{B}(u,v) = ((u,v)) \qquad u,v \in X$$

(3.5) $$\mathcal{A}(u,v) = \int_\Theta uv\omega$$

ω being an element of X such that :

(3.6) $$k_1 \le \omega(t) \le k_2 \qquad k_1, k_2 > 0 \qquad t \in \Theta$$

If we consider on X the topology defined by the norm $[((u,u)) + (u,u)]^{1/2}$, it is easy to see that $\mathcal{B}$, $\mathcal{A}$ are symmetrical, continuous (with respect to each variable) and semi-positive definites.

We now define a numerical integration procedure on X such that :

(3.7) $$\int_{\Theta} u\omega = \sum_{i=1}^{n} \omega(t_i)u(t_i)\, \mu_i^n (t_1,\ldots,t_n) + 0\,(h_n^{(1+\theta)})\, ((u,u))^{1/2} \quad \theta > 0$$

where h_n is given by :

(3.8) $$h_n = \sup_{t\in\Theta} \{ \operatorname{Inf}_{1\le i\le n} ||t-t_i|| \}$$

($||.||$) is the Euclidean norm in R^k)

We introduce an approximation of the bilinear form A , say A_n, defined by

(3.9) $$A_n (u,v) = \sum_{i=1}^{n} u(t_i)\, v(t_i)\, \omega\,(t_i)\, \mu_i^n (t_1,\ldots,t_n)$$

we then have

(3.10) $$| A(u,v) - A_n (u,v) | \le 0\,(h_n^{(1+\theta)})\ |u||v|$$

where $|u| = ((u,u))^{1/2}$

In the following lemma we establish the relation between $n\, D_n^{-1}\Omega_n$ and the bilinear forms B and A_n.

<u>Lemma 3.1</u> Let $\alpha_{1n} \le \alpha_{2n} \le \ldots \le \alpha_{nn}$ be the eigenvalues of the variational problem :

(3.11) $$B(u_n,v) = \alpha_n A_n(u_n,v) \qquad \text{for all } v \in X \qquad |u_n| = 1$$

If u_n is an eigenfunction of (3.11) corresponding to the eigenvalue α_n, th

(i) $u_n \in S_n$

(ii) if $x \in R^n$ is the vector of values of u_n at the knots, and $\bar{D}_n$ is a diagonal matrix with diagonal elements $\omega(t_i)\, \mu_i^n (t_1,\ldots,t_n)$, we have :

(3.12) $$\Omega_n x = \alpha_n \bar{D}_n x$$

and conversely.

<u>Proof</u>

(i) if u_n is a solution of (3.11), we have

$$((u_n , v)) = \alpha_n \sum_{i=1}^{n} u_n (t_i)\, v\,(t_i)\, \omega(t_i)\, \mu_i^n (t_1,\ldots,t_n)$$

$$= \sum_{i=1}^{n} (\alpha_n u_n (t_i)\, \omega\,(t_i)\, \mu_i^n (t_1,\ldots,t_n))\, v\,(t_i)$$

for all $v \in X$

then $u_n \in S_n$ (see definition of S_n).

(ii) S_n being a finite dimensional space, (3.11) is equivalent to :

(3.13) $$B(u_n , \sigma_n^i) = \alpha_n A_n(u_n, \sigma_n^i) \qquad i = 1,2,\ldots,n$$

but $$B(u_n , \sigma_n^i) = \sum_{j=1}^{n} x_j\, \omega_{ij}^n \qquad i = 1,2,\ldots,n$$

and

$$A_n(u_n, \sigma_n^i) = x_i\, \omega(t_i)\, \mu_i^n (t_1,\dots,t_n)$$

we then get :

$$\Omega_n x = \alpha_n \bar{D}_n x$$

The converse part is obvious.

From here on, we will suppose that we have the following important equallity :

(3.14) $$\omega(t_i) = \frac{1}{n\, \mu_i^n (t_1,\dots,t_n)}\, \alpha_i^2 \qquad i = 1,2,\dots,n$$

Typical cases will be described later

With this equality we obtain that $\bar{D}_n = \frac{1}{n} D_n$ and

(3.15) $$\alpha_{in} = \lambda_{in} \qquad i = 1,2,\dots,n$$

Let us call $\lambda_1 \le \lambda_2 \le \dots$ the eigenvalues of the variational problem :

(3.16) $$B(u,v) = \lambda A(u,v) \quad \text{for all } v \in X \qquad |u| = 1$$

We know that A_n approximates A. What can we say on the behaviour of λ_{in} $i = 1,2,\dots,n$? The answer is given by the following theorem.

<u>Theorem 3.2</u> If $\lim_{n\to\infty} h_n = 0$, then $\lim_{n\to\infty} \lambda_{in} = \lambda_i$ for each fixed i

<u>Proof</u> Define $\tilde{B} = B + A$ and consider the problem :

(3.17) $$\tilde{B}(u,v) = \beta\, A(u,v) \quad \text{for all } v \in X \qquad |u| = 1$$

In other words, we translate all the eigenvalues λ_i $i = 1,2,3,\dots$ in order to obtain only non zero eigenvalues , we then have :

a) $\beta_i = \lambda_i + 1 \qquad i = 1,2,\dots$

b) $\tilde{B}(u,u) = |u|^2$

If we define β_{in} $i = 1,2,\dots,n$ as the eigenvalues of the problem :

(3.18) $$B(u_n,v) + A_n(u_n,v) \equiv \tilde{B}_n(u_n,v) = \beta_n A_n(u_n,v) \quad \text{for all } v \in X \quad |u_n| = 1$$

we have :

(3.19) $$\beta_{in} = \lambda_{in} + 1 \qquad i = 1,2,\dots,n$$

Now, consider the error functions :

(3.20) $$\eta_A^n = \sup_{\substack{|u| = 1\\ |v| = 1}} |A(u,v) - A_n(u,v)| \le 0\,(h_n^{1+\theta}) \qquad \theta > 0$$

(3.21) $$\eta_B^n = \sup_{\substack{|u| = 1 \\ |v| = 1}} | \tilde{B}(u,v) - \tilde{B}_n (u,v)| \leq 0\ (h_n^{1+\theta})$$

Now, the hypothesis on h_n implies that :

(3.22) $$\lim_{n\to\infty} \eta_A^n = 0$$

(3.23) $$\lim_{n\to\infty} \eta_B^n = 0$$

The fact that $\tilde{B}$,A are symmetrical implies that the eigenvalues are simple and it is now sufficient to apply a theorem of Fix (cf [8]) to obtain the result :

$$\lim_{n\to\infty} |\lambda_{in} - \lambda_i| = \lim_{n\to\infty} |\beta_{in} - \beta_i| = 0 \quad \text{for each fixed } i$$

This theorem shows the relationship between the eigenvalues associated with a spline function and those of a variational ' continuous ' problem. We are now interested in knowing the behaviour of Tr $(A_n(\tau))$ when n increases. To do this, we need an additional definition.

<u>Definition</u> We will say that $\lambda_1, \lambda_2, \ldots$ <u>increases quickly to infinity</u>, if :

(i) $$\sum_{i\in N} \frac{1}{\lambda_i} < +\infty$$

$$\lambda_i \neq 0$$

(ii) $$\sum_{i=k}^{\infty} \frac{1}{\lambda_i} = 0\ (\lambda_k^{-1})$$

Now we can formulate the main result of this paper :

<u>Theorem 3.3</u> If the eigenvalues $\lambda_1, \lambda_2 \ldots$ increases quickly to infinity and

$$\lim_{n\to\infty} h_n = 0 \quad \text{with } nh_n < c \quad \text{for } n \geq N_0$$

then :

$$\text{Tr } (A_n(\tau)) = \sum_{i=1}^{n} \frac{1}{1+\tau\lambda_i} + \xi_n(\tau)$$

where

$$\lim_{n\to\infty} \xi_n(\tau) = 0 \quad \text{for each } \tau$$

<u>Proof</u> Let us define $R(\tau)$, $R_n(\tau)$ and $\bar{R}_n(\tau)$ by the following variational equalities

$$\tau B(R(\tau)u,v) + A(R(\tau)\ u,v) = A(u,v) \qquad \text{for all } v \in X$$

$$\tau B(R_n(\tau)u,v) + A_n(R_n(\tau)u,v) = A_n(u,v) \quad \text{for all } v \in X$$

$$\tau B(\bar{R}_n(\tau)u,v) + A\ (\bar{R}_n(\tau)\ u,v) = A\ (u,v) \quad \text{for all } v \in S_n \qquad \bar{R}_n(\tau)u\in S_n$$

It is easy to see that $\bar{R}_n(\tau)u$ represents the Raleigh-Ritz-Galerkin approximation of $R(\tau)u$. Then we have :

$$\mathrm{Tr}\,(R(\tau)) = \sum_{i\in N} \frac{1}{1+\tau\lambda_i}$$

$$\mathrm{Tr}\,(R_n(\tau)) = \sum_{i=1}^{n} \frac{1}{1+\tau\lambda_{in}} = \mathrm{Tr}\,(A_n(\tau))$$

$$\mathrm{Tr}\,(\bar{R}_n(\tau)) = \sum_{i=1}^{n} \frac{1}{1+\tau\bar{\lambda}_{in}}$$

where $\bar{\lambda}_{1n} \le \bar{\lambda}_{2n} \le \ldots \le \bar{\lambda}_{nn}$ are the Raleigh-Ritz-Galerkin eigenvalue approximations of $\lambda_1, \lambda_2, \ldots$

$R(\tau)$ being a compact, self adjoint operator, we have the following result (cf [3] , [2])

(3.23) $\lim_{n\to\infty} |\lambda_i - \bar{\lambda}_{in}| = 0$ for each fixed i

(3.24) $\lambda_i \le \bar{\lambda}_{in} \qquad 1 \le i \le n$

Let $\eta > 0$ be a real number. If we choose $J \in N$ such that :

(3.25)
$$\sum_{i=J}^{\infty} \frac{1}{1+\tau\lambda_i} < \eta$$

then we have :

$$\sum_{i=J}^{n} \frac{1}{1+\tau\bar{\lambda}_{in}} \le \sum_{i=J}^{n} \frac{1}{1+\tau\lambda_i} \le \sum_{i=J}^{\infty} \frac{1}{1+\tau\lambda_i} < \eta$$

and we obtain

(3.26)
$$\left|\sum_{i=1}^{n} \frac{1}{1+\tau\lambda_i} - \sum_{i=1}^{n} \frac{1}{1+\tau\bar{\lambda}_{in}}\right| \le \sum_{i=1}^{J} \left|\frac{1}{1+\tau\lambda_i} - \frac{1}{1+\tau\bar{\lambda}_{in}}\right| + 2\eta$$

Now, if we choose $N_0 \in N$ such that $\sum_{i=1}^{J} \left|\frac{1}{1+\tau\lambda_i} - \frac{1}{1+\tau\bar{\lambda}_{in}}\right| < \eta$

for all $n \ge N_0$, we obtain

$$\left|\sum_{i=1}^{n} \frac{1}{1+\tau\lambda_i} - \sum_{i=1}^{n} \frac{1}{1+\tau\bar{\lambda}_{in}}\right| \le 3\,\eta \quad \text{for } n \ge N_0$$

On the other side, we have :

$$\left|\sum_{i=1}^{n} \frac{1}{1+\tau\lambda_{in}} - \sum_{i=1}^{n} \frac{1}{1+\tau\bar{\lambda}_{in}}\right| = |\,\mathrm{Tr}\,(Rn\,(\tau)) - \mathrm{Tr}\,(\bar{R}_n\,(\tau))|$$

$$\le n\ ||R_n\,(\tau) - \bar{R}_n\,(\tau)||$$

since both operators have the same rang S_n. We now apply the theorem of Fix which states that (see [8])

$$|| R_n(\tau) - \bar{R}_n(\tau)|| \leq 0\ (h_n^{1+\theta})$$

and we have :

$$| \mathrm{Tr}\ (Rn(\tau)) - \mathrm{Tr}\ (\bar{R}_n(\tau))| \leq n\ 0\ (h_n^{1+\theta}) = 0\ (h_n^{\theta}) \qquad \theta > 0$$

But θ being a positive number, we can choose $N_0' \geq N_0$ such that $0\ (h_n^{\theta}) < \eta$ and we obtain :

$$\left| \sum_{i=1}^{n} \frac{1}{1+\tau\lambda_i} - \mathrm{Tr}\ (A_n(\tau))\right| \leq 4\eta$$

4 - THE NATURAL POLYNOMIAL SPLINE

In this section, we study the application of the preceding theorems to the case of polynomial splines.

We set $X = H^q [a,b]$, the Sobolev space of real functions defined on [a, whose q - 1 first derivatives are absolutely continuous and it q - th derivative is a square integrable function.

On this space we can define the inner product

(4.1) $$((f,g)) = \int_a^b f^{(q)}(t)\ g^{(q)}(t)\ dt \qquad f,g \in H^q [a,b]$$

(q is an integer $q \geq 1$)

Let $a < t_1 < t_2 < \ldots < t_n < b$ be n points of $[a,b]$ $(n \geq q)$. It is well known (cf [1] , [11] , [13]) that the elements σ of S_n (the space of spline functions) satisfy :

(i) σ is a polynomial of degree q-1 in $[a,t_1]$, $[t_n,b]$

(ii) σ is a polynomial of degree 2q-1 in each interval $[t_i, t_{i+1}]$ $i = 1,2,\ldots,n-1$

(iii) $\sigma^{(j)}$ are continuous $j = 0,1,\ldots,2q-2$

To obtain the theorems on the asymptotic behaviour of the eigenvalues, we must verify the hypothesis (3.1), (3.2), (3.3). The only non trivial hypothesis is (3.3), but applying a classical theorem on Sobolev Spaces we can see that it is satisfied (see Necas [12]).

Now we study the eigenvalues of the 'continuous' problem :

(4.2) $$\int_a^b u^{(q)}(t)\ v^{(q)}(t)\ dt = \lambda \int_a^b u(t)\ v(t)\ \omega(t)\ dt \qquad \text{for all } v \in H^q [a,b]$$

This problem is equivalent to the differential problem :

(4.3)
$$(-1)^q \frac{d^{2q}u}{dt^{2q}} = \lambda\omega u$$
$$u^{(j)}(a) = u^{(j)}(b) = 0 \qquad j = q,\ldots,2q-1$$

In [22] , we have shown that there exist positive constants C_1, C_2 such that

(4.4) $$C_1 i^{2q} \le \lambda_{i+q} \le C_2 i^{2q}$$

But we cannot find an explicit expression for these eigenvalues in the general case. For this reason, we consider now two cases.

4.1 The Case of Equally Spaced Data and Constant Standard Deviation

Let us suppose that the knots $t_1,\dots,t_n$ are given by :

$$t_i = \frac{2i-1}{n}(b-a) + a \qquad i = 1,2,\dots,n$$

and that the standard deviation of each random number ε_i is equal to v.

In this case, the integration formula can be taken to be :

(4.5) $$\int_a^b g(t)\,dt \cong \frac{b-a}{n} \sum_{i=1}^{n} g(t_i)$$

and the error bound is given by (cf : [ ])

(4.6) $$\left| \int_a^b g(t)\,dt - \frac{b-a}{n} \sum_{i=1}^{n} g(t_i) \right| \le c \left(\frac{b-a}{n}\right)^2 \left| g \right|$$

The weight function becomes :

$$\omega(t_i) = \frac{1}{n\,\mu_i^n(t_1,\dots,t_n)}\,\alpha_i^2$$

$$= \frac{1}{n\left(\frac{b-a}{n}\right)} = \frac{1}{b-a} \qquad i = 1,2,\dots,n$$

and

(4.7) $$\omega(t) = \frac{1}{b-a} \qquad t \in [a,b]$$

Finally, the differential problem (4.3) is :

(4.8)

$$(-1)^q \frac{d^{2q}u}{dt^{2q}} = \frac{1}{b-a}\,\lambda u$$

$$u^{(j)}(a) = u^{(j)}(b) = 0 \qquad j = q,\dots,2q-1$$

The solutions to (4.8) are well known in two cases : q = 1,2. The case q = 1 is not very useful (linear splines), but the case q = 2 is one of the more important cases (cubic splines). We have also calculated the solutions in the case q = 3 (quintic splines). In [21] , [22] we tabulate the eigenvalues of this problem in a normalized interval.

4.2 The General Case

As we have already stated, we cannot calculate in a simple way, the eigenvalues of the continuous problem with general weight function. To solve this problem, we propose now a practical method :

Let $m > n$ and consider an approximation of the continuous problem by a finite difference method. The convergence of such methods (for the eigenvalue probler has been studied by Kreiss [10] . We suppose that m is large and we consider the first n eigenvalues of the finite difference approximation, as the exact ones and we utili them to calculate the trace of $A_n(\tau)$. It might be possible to show that this approximation is convergent, but we will not do it here.

We have then reduced our eigenvalue problem with a full symmetrical matrix, to the problem :

$$\frac{1}{(m+1)^{2q}} \quad Fx = \theta\,\bar{D}x \qquad\qquad m >> n \tag{4.9}$$

where F is the discretisation matrix, and $\bar{D}$ is a diagonal matrix containing the values of ω at the mesh points.

This problem is equivalent to the following one :

$$\frac{1}{(m+1)^{2q}} \quad \bar{D}^{-1/2} F \bar{D}^{-1/2} v = \theta\, v \qquad\qquad v \in R^m \tag{4.10}$$

The matrix $\bar{D}^{-1/2} F \bar{D}^{-1/2}$ is a symmetrical band matrix (with $2q + 1$ bands). We can reduce it to a tridiagonal form by the Schwarz method (see [16]) in $0\ (m^2\ )$ operations. If $m = 2n$ we have an $0\ (n^2\ )$ method to calculate the required eigenvalue

This method has been tested successfully in the cubic spline case. For thi case, the matrix F is given by :

$$F = \begin{bmatrix} -6 & 4 & -1 & & & \\ 4 & -6 & 4 & -1 & & \\ -1 & 4 & -6 & 4 & -1 & \\ & & & & & \\ & & & & & \\ & & & -1 & 4 & -6 \end{bmatrix} \tag{4.11}$$

To obtain $\bar{D}$, we need an expression of ω , but we suppose that :

$$\omega(t_i) \quad = \frac{1}{n\,\mu_i(t_1,\ldots,t_n)}\ \alpha_i^2 \qquad\qquad i = 1,2,\ldots,n$$

We have observed that it is sufficient to replace ω by its interpolating first order spline, to obtain very good eigenvalue approximation (at least for our calculations). In the next section we will show a set of numerical results and a run-time table.

5 - NUMERICAL EXAMPLES

In this section, we show a few numerical examples made on artificial data, we consider a function f defined on $[0,1]$ and calculate their values at points

$t_1, t_2, \ldots, t_n$. Next, we generate pseudo-random numbers ε_i with $N(0,v^2)$ distribution and add these numbers to the values of the function. These data are smoothed by the Cross-Validation Method and the result is compared to f.

5.1 The Equally Spaced Data Case

Here, we consider n equally spaced points in [0,1] and a function f given by :

$$f(t) = \pi \cos \pi t + \frac{3\pi}{2} \cos 3\pi t$$

The standard deviation of the pseudo-random numbers is v = 0.15, and the test is made for n = 50, 100, 200, 400. The results are shown in figures 1 to 4, where we plot for each case, the date (x), the smoothed spline (s) and the function (F). We observe the remarkable convergence properties of the solution and think it might be possible to show this result.

Further examples and the comparison with other methods is given in [22]. A detailed description of the FORTRAN programs developed to carry out the calculations is given in [20].

The run times (*) are given in table 1 as a function of n (in seconds). The first column gives the time for evaluating $\lambda_{1n}, \lambda_{2n}, \ldots, \lambda_{nn}$. The second column represents the time for estimating the optimal parameter, and the third column shows the total run time.

Table 1

n	E.C	O.P	T.T
50	0.0031	0.7962	0.7993
100	0.0066	1.1310	1.1376
150	0.0098	1.7466	1.7564
200	0.0134	2.4569	2.4703
250	0.0215	3.1580	3.1795
300	0.0193	3.7720	3.7913
350	0.0238	4.4208	4.4446
400	0.0269	4.8706	4.8975

It is easy to see that run times increase linearly with n (approximately) and we have :

$$\text{run time} \cong 0.012\, n$$

The only practical problem is the storage :we need a 4n matrix and other vectors. So, in the next section we develop a method for the case of large number of points.

(*) On an IBM 360/67 computer.

CROSS - VALIDATION

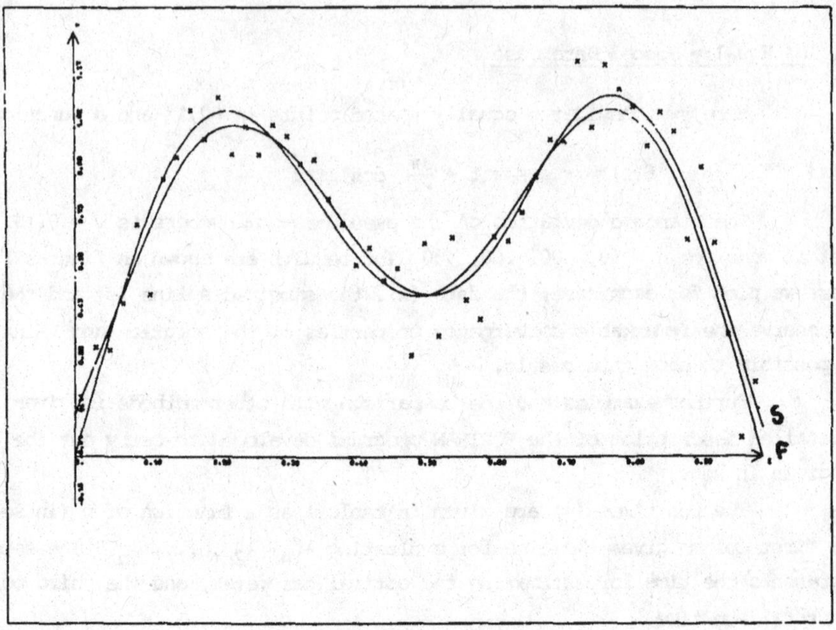

FIGURE 1: 50 POINTS

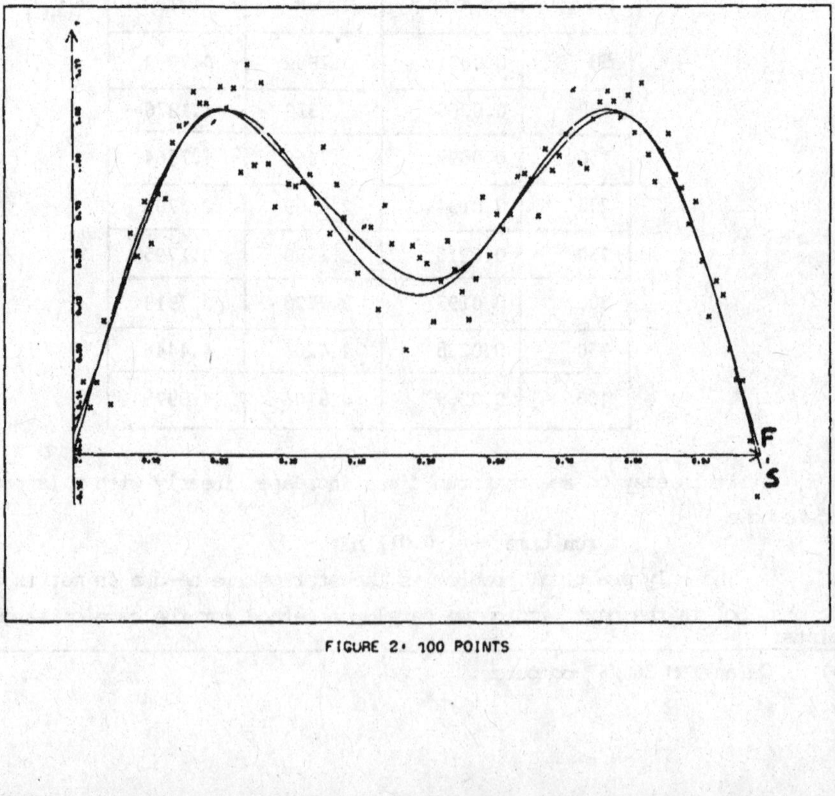

FIGURE 2: 100 POINTS

CROSS - VALIDATION

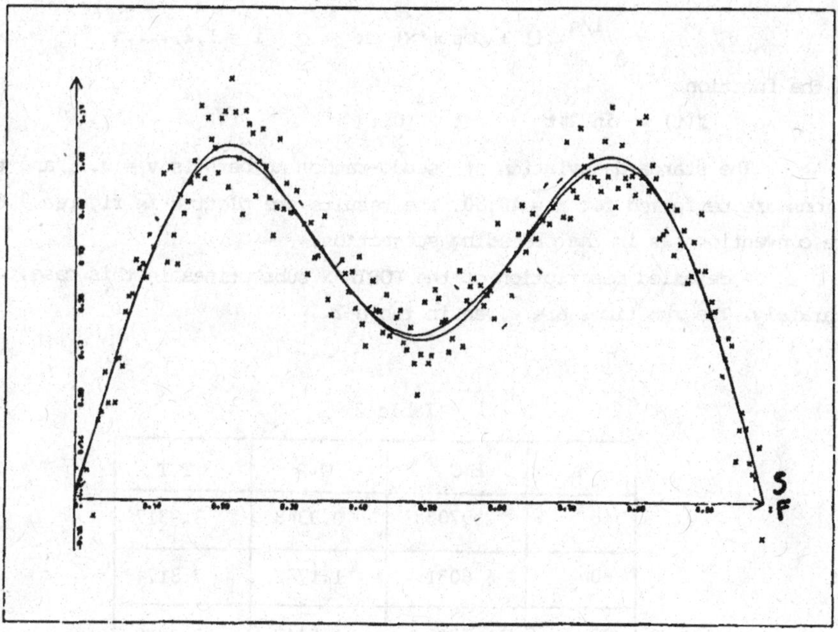

FIGURE 3: 200 POINTS

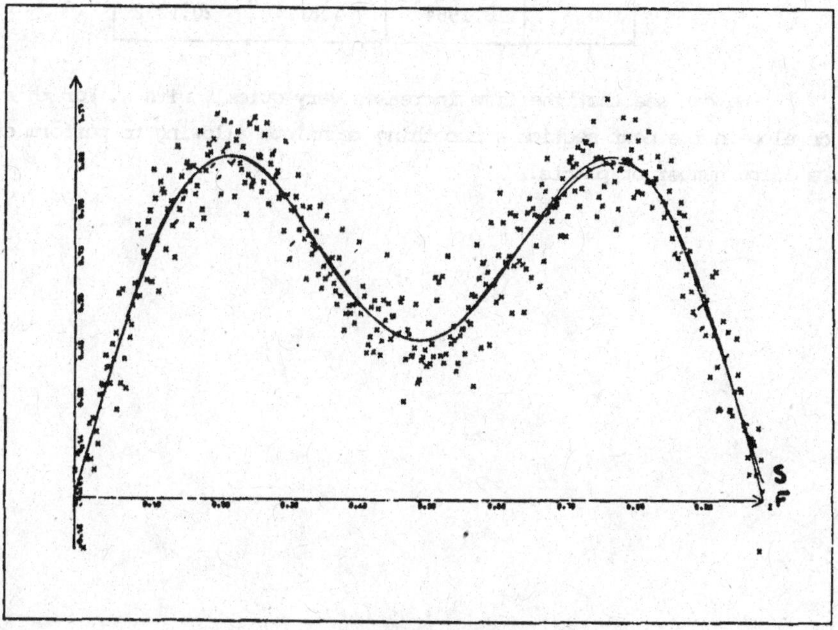

FIGURE 4: 400 POINTS

UTRERAS LE 05.01.79 A 12.31.57 DESSIN NO 003

5.2 The General Case

In this subsection, we consider n points in $[0,1]$ generated by :

$$t_i = \int_0^{i/n} (1 + \cos \pi x)\, dx \qquad i = 1,2,\ldots,n$$

and the function

$$f(t) = \sin 2\pi t \qquad t \in [0,1]$$

The standard deviation of pseudo-random numbers is $v = 0.1$, and the calculations are performed for $n = 40,80$. The results are plotted in figures 5,6 with the same conventions as in the preceding subsection.

A detailed description of the FORTRAN subroutines in this case, will appear separately. The run times are given in table 2.

Table 2

n	E.C	O.P	T.T
40	2.9703	0.9348	3.9317
60	6.6031	1.1742	7.8174
80	11.7702	1.5447	13.3709
100	18.1964	1.8627	20.1272

We can see that the time increases very quickly with n. For this reason, we develop in the next section a smoothing technique allowing to perform calculations for a large number of points.

CROSS - VALIDATION

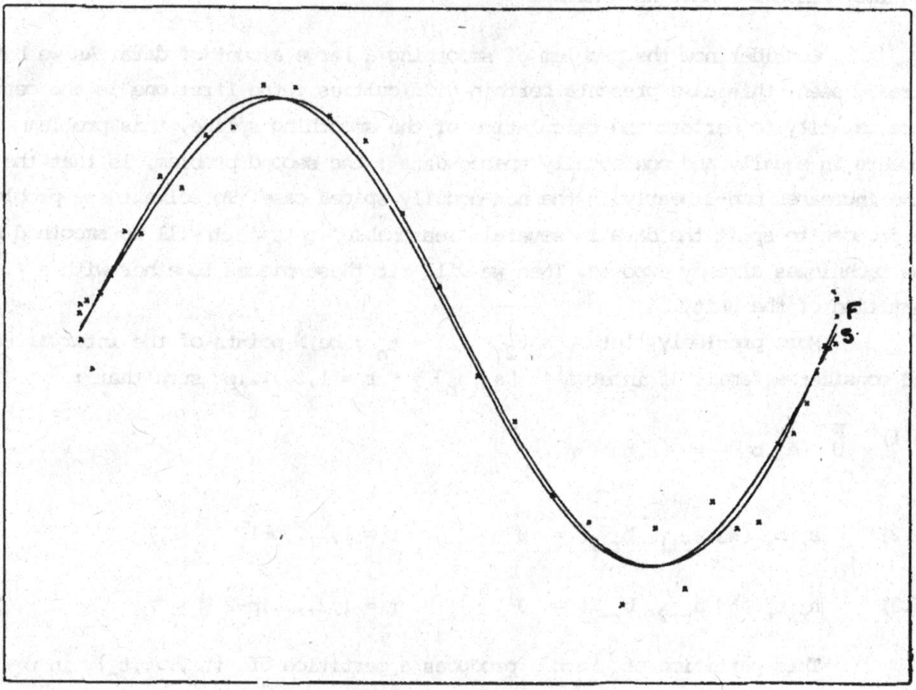

Exemple 5 : 40 points

CROSS - VALIDATION

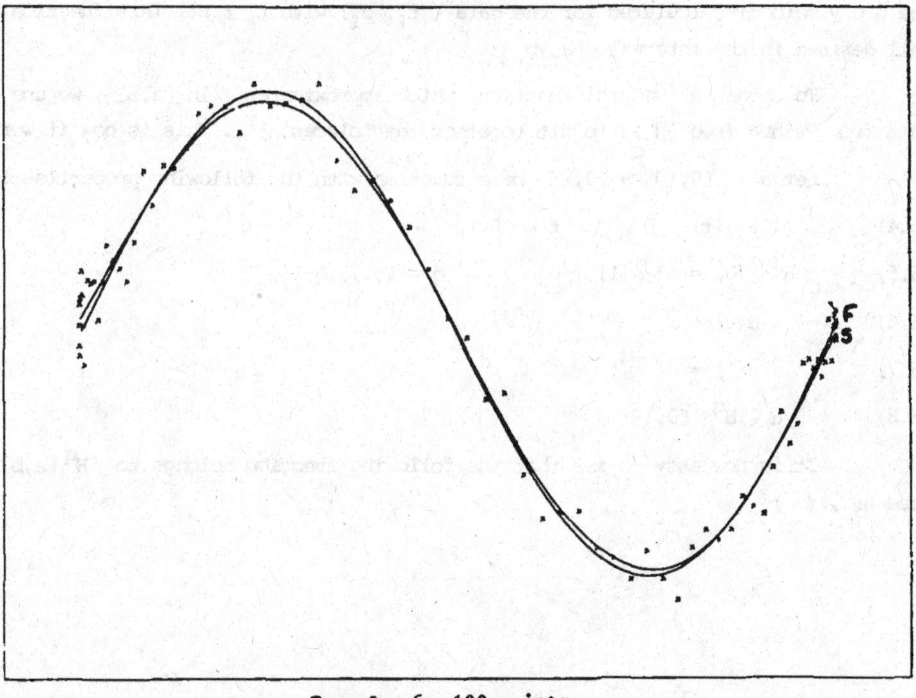

Exemple 6 : 100 points

6 - THE PIECEWISE SMOOTHING METHOD

Consider now the problem of smoothing a large amount of data. As we have already seen, this case presents certain difficulties : the first one is the central core capacity to perform the calculation of the smoothing spline, this problem appears in equally and non equally spaced data ; the second problem, is that the run time increases non-linearly in the non-equally spaced case. To solve these problems, we propose to split the data in several 'reasonable' sets which will be smoothed by the techniques already exposed. Then we will fit these pieces together with a partition of the unity.

More precisely, let $t_1 < t_2 < \dots < t_n$; be N points of the interval [a,b] and consider a family of intervals $[a_r,b_r]$ $r = 1,2,\dots,p$ such that :

(6.1) $$\bigcup_{r=1}^{p} [a_r,b_r] = [a,b]$$

(6.2) $$]a_r,b_r[\cap]a_{r+1}, b_{r+1}[\neq \emptyset \qquad r = 1,\dots,p-1$$

(6.3) $$]a_r,b_r[\cap]a_{r+2}, b_{r+2}[= \emptyset \qquad r = 1,2,\dots,p-2$$

This partition of [a,b] produces a partition of $\{t_1,\dots,t_n\}$ in p subsets :

$$P^r = \{t \in \{t_1,\dots,t_n\} \ / \ t \in [a_r,b_r]\} \qquad r = 1,2,\dots,p$$

Let $\tilde{\sigma}^r$ be the smoothing spline of order q and parameter τ^r (chosen by the G C V method) calculated for the data (t_ℓ, z_ℓ) with $t_\ell \in P^r$. This function is well defined in the interval $[a_r,b_r]$.

To obtain a 'smooth' function, which approximates f in [a,b] , we use an idea from Paihua (see [13]), to fit together the 'pieces' $\tilde{\sigma}^r$. This is how it works :

Let $\alpha : [0,1] \to [0,1]$ be a function with the following properties :

(6.4) $$1 \geq \alpha(t) \geq 0 \qquad t \in [0,1]$$

(6.5) $$\alpha^{(j)}(0) = \alpha^{(j)}(1) = 0 \qquad j = 1,\dots,q-1$$

(6.6) $$\alpha(0) = 1$$

(6.7) $$\alpha(1) = 0$$

(6.8) $$\alpha \in H^q\,[0,1]$$

It is now easy to see that the following function belongs to $H^q[a,b]$ (see [13] , [22])

$$\sigma(t) = \begin{cases} \hat{\sigma}^r(t) & \text{for } t \in [a,b] - \bigcup_{r=1}^{p-1} (]a_r,b_r[\cap]a_{r+1},b_{r+1}[) \\ \alpha\left(\frac{t-a_r}{b_r-a_r}\right) \hat{\sigma}^r(t) + \left(1 - \alpha\left(\frac{t-a_r}{b_r-a_r}\right)\right) \hat{\sigma}^{r+1}(t) & \text{for } t \in [a_{r+1},b_{r+1}] \cap [a_r,b_r] \end{cases}$$

To analyse the value of this approximation, it is important to remember that f each $\hat{\sigma}^r$ was an interpolating function of f, σ would also be an interpolating unction of f. It is reasonable to conclude that if each $\hat{\sigma}^r$ is a 'good' approximation f f, σ will also be a 'good' approximation of f.

The choice of the intervals $[a_r,b_r]$ $r = 1,\ldots,p$; and the choice of the unction α still remain. The second problem is easy to solve if we impose that α e a polynomial of minimal degree, we obtain :

6.10) $\alpha(t) = (2t + 1)(1-t)^2$ for $q = 2$

6.11) $\alpha(t) = (6t^2 + 3t + 1)(1-t)^3$ for $q = 3$

When choosing the partition $[a_r,b_r]$ $r = 1,\ldots,p$ we must remember that the umber of data points in $[a_r,b_r]$ must be reasonable, that is, not too great, in order o avoid computer limitations, and not too small, in order to approximate f as well s possible. The main problem is the choice of the number of points in each intersection. n [22] , we carry out a large set of numerical tests and conclude that the number of oints in each intersection must be greater than 25 % of the points in the corres- onding intervals for equally spaced data, and 40 % for non equally spaced data.

To illustrate the results of this method, we present an example of artifi- ial data generated as in section 5. We use a test function $f(x) = \sin x$, the data re equally spaced, the distance between mesh points is 0.1 an the standard deviation s 0.5. All partitions contain 100 points. In figure 7 the number of points in the ntersection is 40 and in figure 8 it is 20. The data is represented by a broken line nd F and S represent the function and the solution respectively.

PIECEWISE SMOOTHING TECHNIQUE

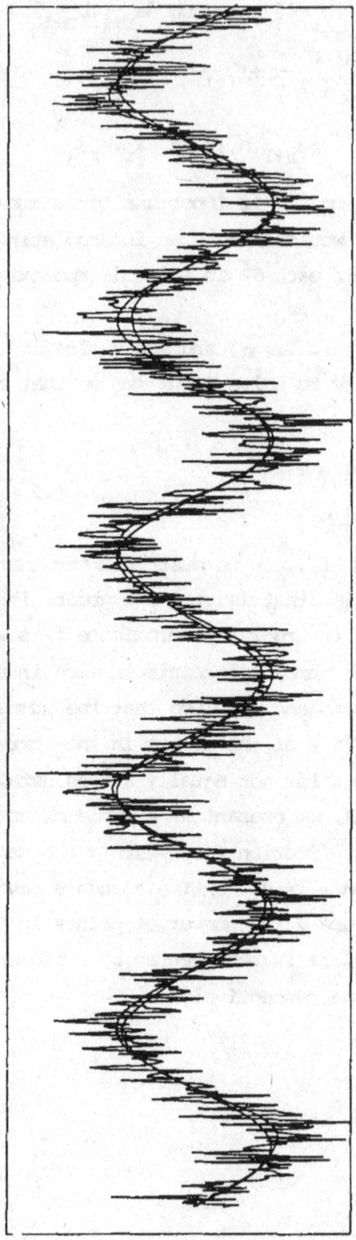

Exemple 7 : n = 100 ; k = 40 ; v = 0,5

COLLE LE 01/02/79 A 22:11:14 D

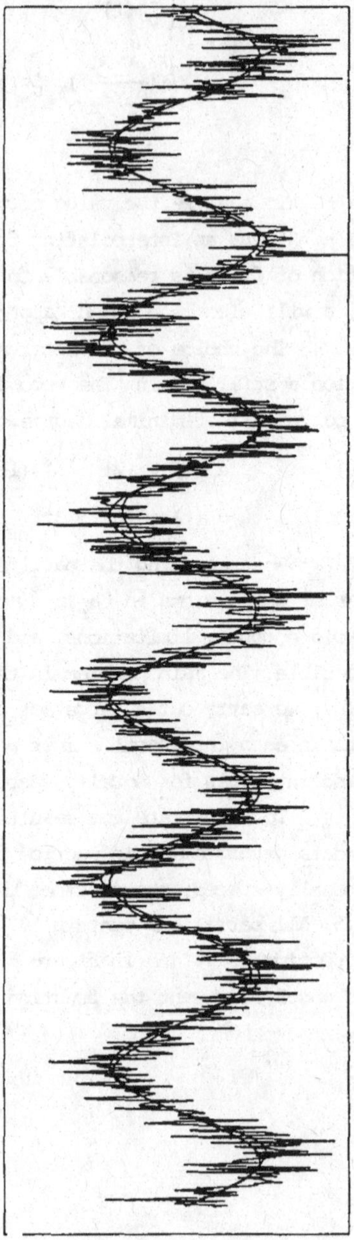

Exemple 8 : n = 100 ; k = 20 ; v = 0,5

COLLE LE 01/02/79 A 22:11:32

7 - BIVARIATE SPLINES ON A BOUNDED SET

Our aim in this section is to show how results of section 3 can be applied in the two dimensional case.

Let Θ be a bounded open set of R with Lipschitz boundary (see Necas [12]). We will study two types of spline functions defined on Θ : The Thin Plate Splines with zero Boundary Condition, and the Thin Plate Splines with free boundary condition.

7.1 The Thin Plate Splines With Zero Boundary Condition

Let $X = H_0^2(\Theta)$ be defined by :

(7.1) $$H_0^2(\Theta) = \{f : \Theta \to R \ / \quad D^{i,j}f \in L^2(\Theta) \qquad i+j \leq 2$$
$$f\Big|_{\partial\Theta} = 0 \quad \frac{\partial f}{\partial \vec{n}}\Big|_{\partial\Theta} = 0\ \}$$

Where $\frac{\partial f}{\partial \vec{n}}$ is the normal derivative of f, the derivatives being taken in the sense of distributions.

On $H_0^2(\Theta) \times H_0^2(\Theta)$, we consider the bilinear form :

(7.2) $$((u,v)) = \int_\Theta \sum_{i+j=2} (D^{i,j}u)\ (D^{i,j}v)$$

It is well known that ((u,v)) represents (in a first order approximation), the flexion energy of a thin plate. The boundary conditions are those of the clamped plate (see Courant & Hilbert [4])

Denote $|u| = ((u,u))^{1/2}$. For this semi-norm, X is a Hilbert space (and $|.|$, a norm). We can easily verify the hypotheses of section 2 and obtain that there is one and only one solution to the problem :

(7.3) $$\text{Minimize } \{\ ((u,u)) = \sum_{i+j=2} \int_\Theta (D^{i,j}u)^2\ \}$$
$$u \in H_0^2(\Theta)$$
$$u(t_i) = z_i \qquad i = 1,2,\ldots,n$$

The points $t_1,t_2,\ldots,t_n$ are supposed to be distinct and not on a straight line.

So, the eigenvalue problem associated with this type of spline functions, becomes :

(7.4) $$\sum_{i+j=2} \int_\Theta (D^{i,j}u)\ (D^{i,j}v) = \lambda \int_\Theta \omega uv \qquad \text{for all } v \quad H_0^2(\Theta)$$
$$u \in H_0^2(\Theta)$$
$$|u| = 1$$

This problem is equivalent to the following differential problem (see Courant & Hilbert [4] , Thomman [17])

(7.5) $$\Delta^2 u = \lambda\omega u$$

$$u \Big|_{\partial\Theta} = 0$$

$$\frac{\partial u}{\partial n} \Big|_{\partial\Theta} = 0$$

where Δ is the Laplacian operator.

Unfortunately, we cannot obtain a simple analytic expression of the eigenvalues of this problem for general Θ ,ω . The only case which is easy to solve, is the circular boundary case (and $\omega \equiv 1$) where the eigenvalues are calculated as the fourth power of the roots of the following equation :

(7.6) $$\frac{J_n'(x)}{J_n(x)} = \frac{i\,J_n'(ix)}{J_n(ix)}$$

where J_n is the n-th Bessel function (see Courant & Hilbert [4])

This is not surprising, as the reproducing kernel of $H_0^2(\Theta)$ is only known for this case, too.

7.2 The Thin Plate Splines With Free Boundary Condition

Another type of bivariate spline functions has been studied by Thomman [17] Their definition is essentially the same as in the preceding subsection, the only difference is the minimization space.

Let X be the Sobolev Space $H^2(\Theta)$ and define on $H^2(\Theta) \times H^2(\Theta)$ the same semi-norm as in the preceding subsection.

It can be easily verified that the Null space of ((.,.)) is the set of polynomials of degree 1 and that X/N is a Hibert Space (see Necas [12]). All the other hypotheses are easily verified.

Then, the associated eigenvalue problem becomes :

$$\int_\Theta \sum_{i+j=2} (D^{i,j}u)\,(D^{i,j}v) = \mu \int_\Theta \omega u v$$

$$u \in H^2(\Theta) \quad |u| = 1 \qquad \text{for all } v \in H^2(\Theta)$$

This problem is equivalent to an eigenvalue problem associated with the bilaplacian operator, but the boundary conditions are very different from the preceding case (see Thomman [17]) The eigenvalues and the kernel are known only in the case of circular boundary as in the preceding subsection.

8 - THIN PLATE SPLINES IN THE WHOLE PLANE

In the preceding section, we have presented the 'thin plate' splines on a bounded domain, and we have said that the kernel function allowing us to characterize the Spline Functions, can be calculated only in the circular boundary case. In this case, the kernel is expressed as a series containing Bessel functions and is very hard to calculate. Indeed, the difficulty comes from the boundary conditions

For this reason Duchon [7] has introduced a new type of splines defined over all R^2, avoiding the boundary problems and allowing a simple characterization, very useful for numerical calculations. These splines are defined as follows :

Let $D^{-2}L^2(R^2)$ be the vector space of distributions ϕ on R^2 such that $D^{i,j}\phi$ belongs to $L^2(R^2)$ for each i,j such that i+j = 2. It is well known (see e.g. Duchon [7] , Schwartz [15]) that $D^{-2}L^2(R^2)$ is a space of continuous functions.

On $D^{-2}L^2(R^2)$ we can define the following semi-inner product :

(8.1) $$((u,v)) = \sum_{i+j=2} \int_{R^2} (D^{i,j}u)\,(D^{i,j}v)$$

and the associated semi-norm.

It is easy to see that the Null space of this semi-norm is the set of Polynomials of degree 1. And Duchon has proved that $D^{-2}L^2(R^2)/N$ is a Hilbert space. So given $t_1, t_2,\ldots,t_n \in R^2$ $z_1,\ldots,z_n \in R$ such that all the points are not on a straight line, there exists a unique solution to the problem :

(8.2) Minimize $$\{\tau \sum_{i+j=2} \int_{R^2} (D^{i,j}u)^2 + \frac{1}{n}\sum_{i=1}^{n} \alpha_i^2 (z_i - u(t_i))^2\}$$

$u \in D^{-2}L^2(R^2)$

This solution is called the smoothing spline of parameter τ .

The most important result for calculations is that a kernel of $D^{-2}L^2(R^2)$ with the topology of this semi-norm, can be found to be :

(8.3) $$K(t,s) = \frac{1}{4\pi}\,|t-s|^2 \operatorname{Log}|t-s|^2 \qquad t,s \in R^2$$

Paihua [13] has developed a set of algorithms allowing to perform numerical calculations with these splines and FORTRAN subroutines are available (see [14]).

The asymptotic behaviour of the associated eigenvalues is not given by the theorems of the preceding sections, but we think it is associated with an eigenvalue problem with the bi-laplacian operator.

In order to see how smoothing behaves in bivariate function approximation, we have written a set of FORTRAN subroutines for calculating the smoothing parameter by G.C.V., in this case (Thin Plate Splines) The associated eigenvalues are calculated by the QR method, and the optimal parameter is calculated by global search. Paihua's program is used to calculate the smoothing for a given parameter. A table of run times is given below. (Table 3)

It is important to see that the calculation of a smoothing spline with a given parameter is performed in $O(n^3)$ operations, not in $O(n)$ as in one dimensional problems, so that the eigenvalue calculations do not increase the cost too much, as for one dimensional problems.

In figures 9,10,11 we present an illustrating numerical example. In domain $[0,2.3] \times [0,1.5]$ we choose a 100 random points $t_1,\ldots,t_{100}$. We generate artificial data adding pseudo-random normal numbers to the values of the function

$$f(x,y) = \sin\left(\left(x-\frac{1}{2}\right)^2 + \left(y-\frac{1}{2}\right)^2\right)$$

In figure 9 we plot level curves of an interpolating spline to the data. In figure 10, we plot the same levels for the smoothed data, and in figure 11 we plot the true function. We observe that even with a few points (a hundred points in $[0,2.3] \times [0,1.5]$) the result is remarkably good.

There are two problems to perform these calculations : the central core capacity, and the run time (see table 3). For this reason in the next section we develop a piecewise calculation technique inspired by Paihua [13].

Table 3

n	E.C	O.P	T.T
40	4.2370	8.5337	12.7707
60	12.5260	27.6989	40.2249
80	28.4075	59.1651	87.5726
100	53.9608	103.8002	157.7610

We obtain the approximate relation :

$$\text{run time} \cong 0.18 \times 10^{-3} n^3$$

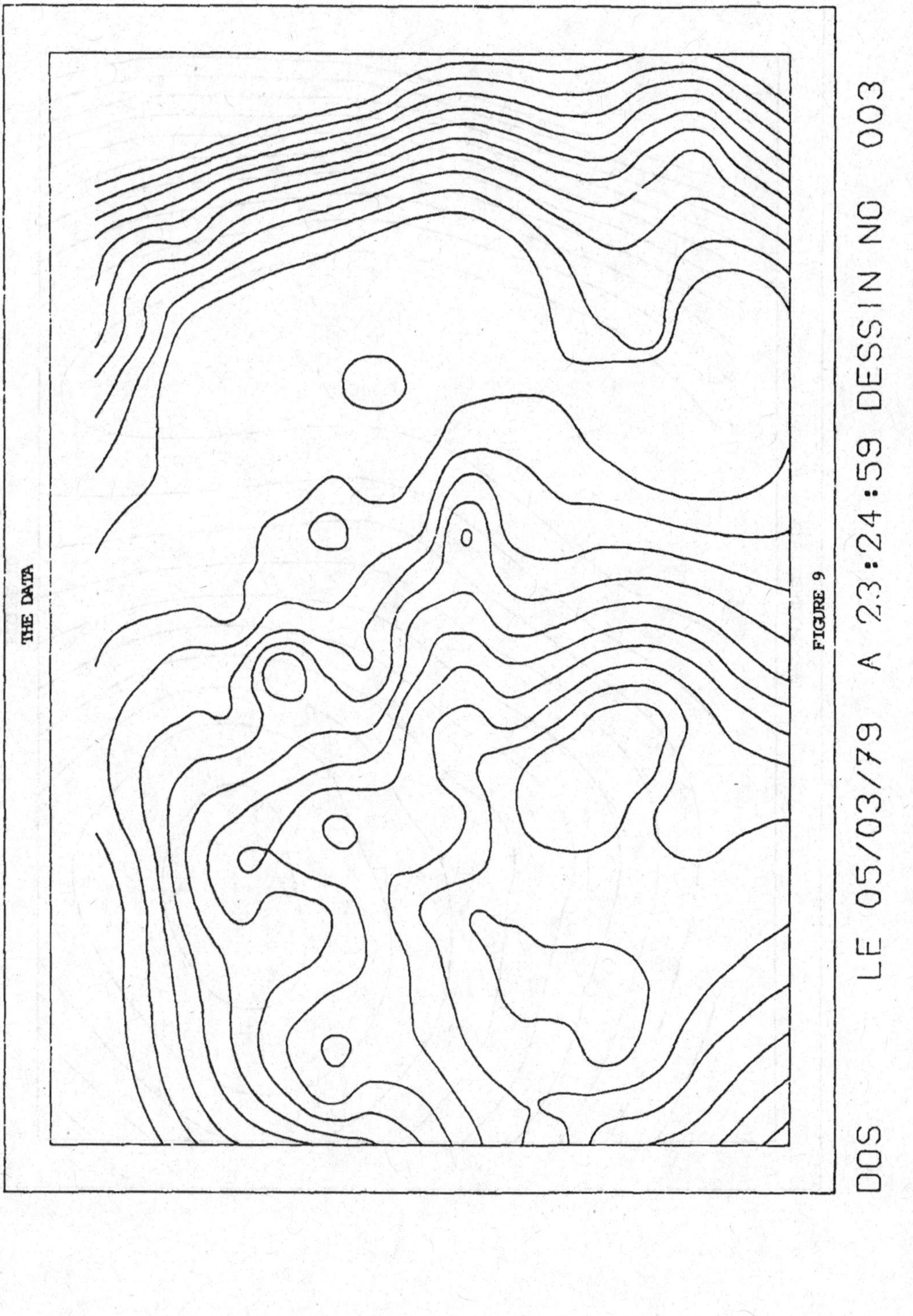

FIGURE 9

THE SMOOTHED DATA

FIGURE 10

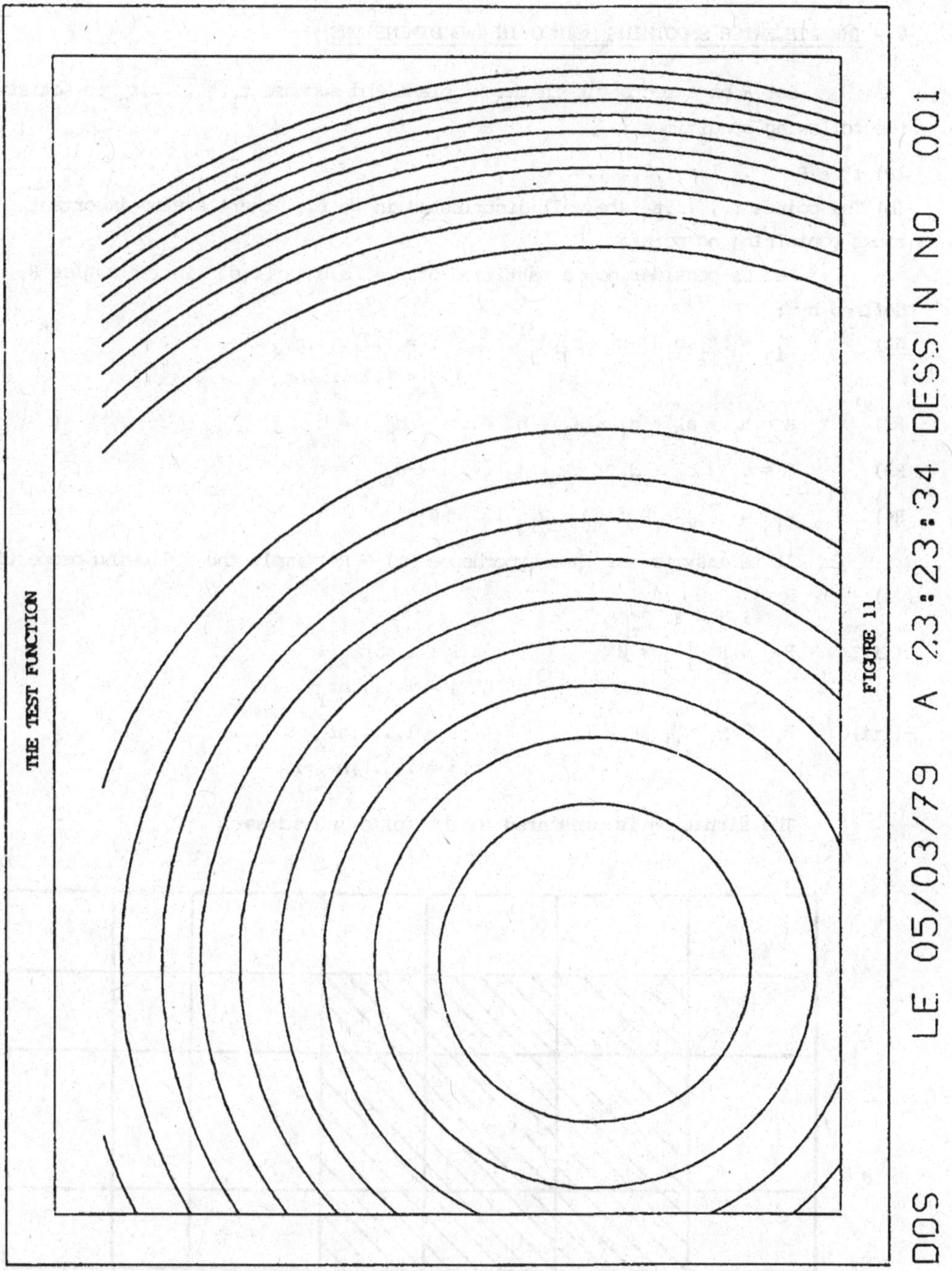

FIGURE 11

9 - THE PIECEWISE SMOOTHING METHOD IN TWO DIMENSIONS

Let R be a rectangle in the R^2 plane and suppose $t_1, t_2, \ldots, t_n$ to satisfy the following conditions :

(a) $t^i \in R \qquad i = 1,2,\ldots,n$

(b) The points $t_1,\ldots,t_n$ are well distributed in R, i.e. there are no imoortant zones containing no points.

Let us consider now a partition of $R = [a,b] \times [c,d]$ in rectangles R_{ij} defined by :

R1) $\quad R_{ij} = [a_i,b_i] \times [c_j,d_j] \qquad i = 1,2,\ldots,nr_x$

$\qquad\qquad j = 1,2,\ldots,nr_x$

R2) $\quad a = a_1 < a_2 < b_1 < a_3 < b_2 < \ldots < b_{nr_x} = b$

R3) $\quad c = c_1 < c_2 < d_1 < c_3 < d_2 < \ldots < d_{nr_y} = d$

R4) $\quad P_{ij} = \{t_k \in R \ / \ t_k \in R_{ij}\} \neq \emptyset$

It is easy to see that hypotheses R1) → R4) imply the following properties

(i) $\quad R = \bigcup_{i,j=1} R_{ij}$

(ii) $\quad R_{ij} \cap R_{i+1,j} \neq \emptyset \qquad 1 \le i \le nr_x-1$

$\qquad\qquad j = 1,\ldots,nr_y$

(iii) $\quad R_{ij} \cap R_{i,j+1} \neq \emptyset \qquad i = 1,\ldots,nr_x$

$\qquad\qquad j = 1,\ldots,nr_y-1$

The situation is presented in the following scheme :

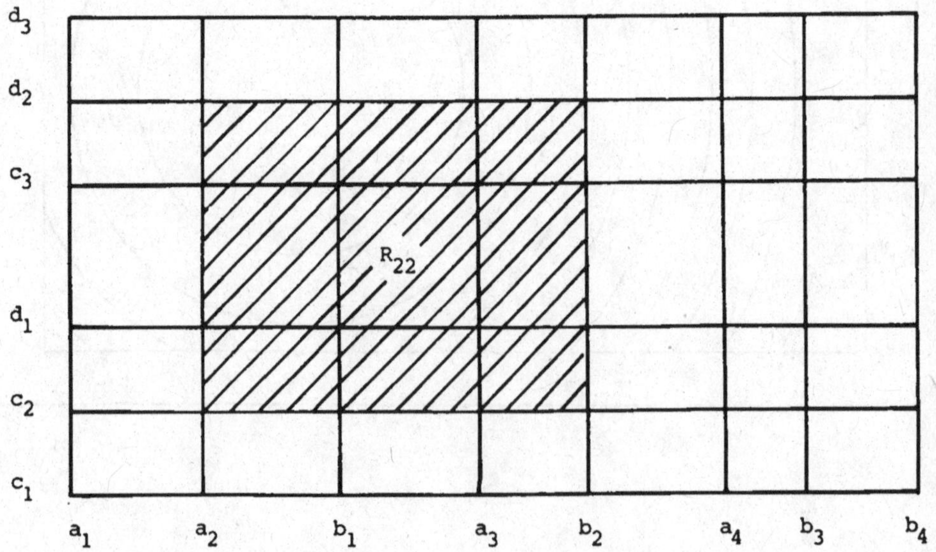

Let $\hat{\sigma}_{ij}$ be the smoothing 'thin plate' spline (with optimal parameter chosen by G C V) calculated on the data contained in the rectangle R_{ij}. This way, we have several 'edges' of smoothed surfaces that do not coincide in the intersection of the definition domain. Now, we proceed to fitting the pieces together with the same technique as in section 6. This idea has already been proposed by Paihua [13] in the interpolation case.

It is very important to note that in the definition of the 'local' splines $\hat{\sigma}_{ij}$ and $\hat{\sigma}_{i+1,j}$ we use the points of the intersection $P_{ij} \cap P_{i+1,j}$ (the same occurs for $\hat{\sigma}_{i,j+1}$ and $\hat{\sigma}_{ij}$ with the points of the set $P_{ij} \cap P_{i,j+1}$, and so on) The fitting will strongly depend on the choice of the size of these intersections.

In [13] , [14] , we show that the function σ defined below is in $C^1(\mathbb{R}^2)$:

$$
(9.1)\quad \sigma(x,y) = \begin{cases}
\hat{\sigma}_{ij}(x,y) \qquad (x,y) \in R_{ij} \quad [R_{ij-1} \cup R_{ij+1} \cup R_{i+1j} \cup R_{i-1j}] \\[2ex]
\alpha\left(\dfrac{x-a_{i+1}}{b_i-a_{i+1}}\right)\hat{\sigma}_{ij}(x,y) + \left(1-\alpha\left(\dfrac{x-a_{i+1}}{b_i-a_{i+1}}\right)\right)\hat{\sigma}_{i+1j}(x,y) \\[1ex]
\qquad (x,y) \in R_{ij} \cap R_{i+1j} - (R_{ij+1} \cup R_{ij-1}) \\[2ex]
\alpha\left(\dfrac{y-c_{j+1}}{d_j-c_{j+1}}\right)\hat{\sigma}_{ij}(x,y) + \left(1-\alpha\left(\dfrac{y-c_{j+1}}{d_j-c_{j+1}}\right)\right)\hat{\sigma}_{i,j+1}(x,y) \\[1ex]
\qquad (x,y) \in R_{ij} \cap R_{i,j+1} - (R_{i+1j} \cup R_{i-1j}) \\[2ex]
\alpha\left(\dfrac{x-a_{i+1}}{b_i-a_{i+1}}\right)\left[\alpha\left(\dfrac{y-c_{j+1}}{d_j-c_{j+1}}\right)\hat{\sigma}_{ij}(x,y) + \left(1-\alpha\left(\dfrac{y-c_{j+1}}{d_j-c_{j+1}}\right)\right)\hat{\sigma}_{ij+1}(x,y)\right] + \\[1ex]
+\left(1-\alpha\left(\dfrac{x-a_{i+1}}{b_i-a_{i+1}}\right)\right)\left[\alpha\left(\dfrac{y-c_{j+1}}{d_j-c_{j+1}}\right)\hat{\sigma}_{i+1j}(x,y) + \left(1-\alpha\left(\dfrac{y-c_{j+1}}{d_j-c_{j+1}}\right)\right)\cdot \right. \\[1ex]
\qquad \left. \cdot\, \sigma_{i+1,j+1}(x,y)\right] \\[1ex]
\qquad (x,y) \in R_{ij} \cap R_{i+1j} \cap R_{i,j+1} \cap R_{i+1,j+1}
\end{cases}
$$

where denotes the function :

$$(9.2)\qquad \alpha(t) = (t-1)^2\,(2t+1) \qquad t \in [0,1]$$

The programs necessary to evaluate σ (and calculate all its parameters) are easily obtained from those already written for the interpolation problem (see [14]) These programs will appear in the near future.

To illustrate the behaviour of the method, we have chosen the same example as in the preceding section, but in this case, we have 700 points. The partition used is presented in figure 12. The other three figures are obtained in the same way as in preceding section.

The run time was 1633 secs, the same problem in direct calculation would have taken 61740 secs, that is to say 38 times our run time. The advantage of this piecewise calculation is obvious.

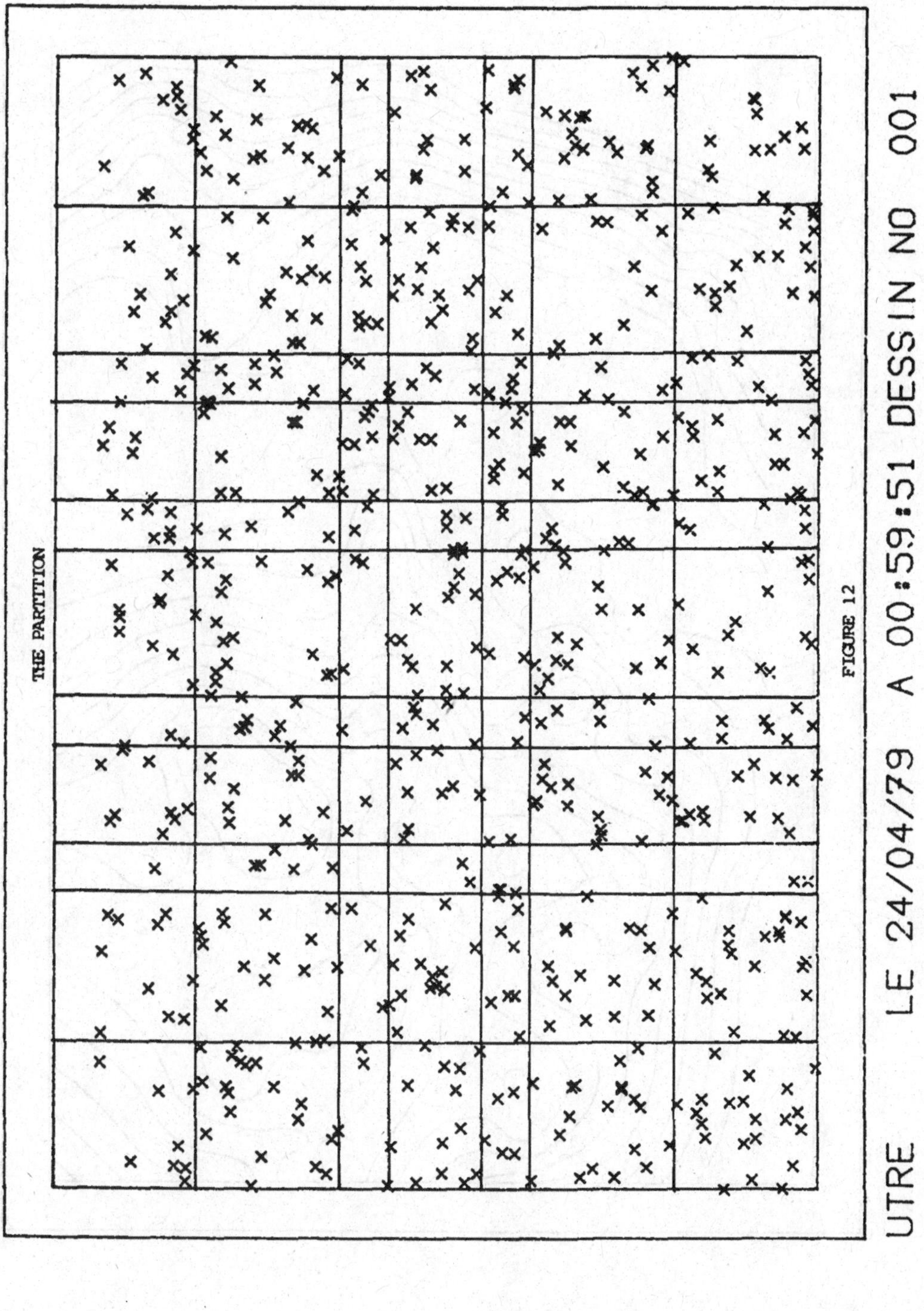

FIGURE 12

FIGURE 13

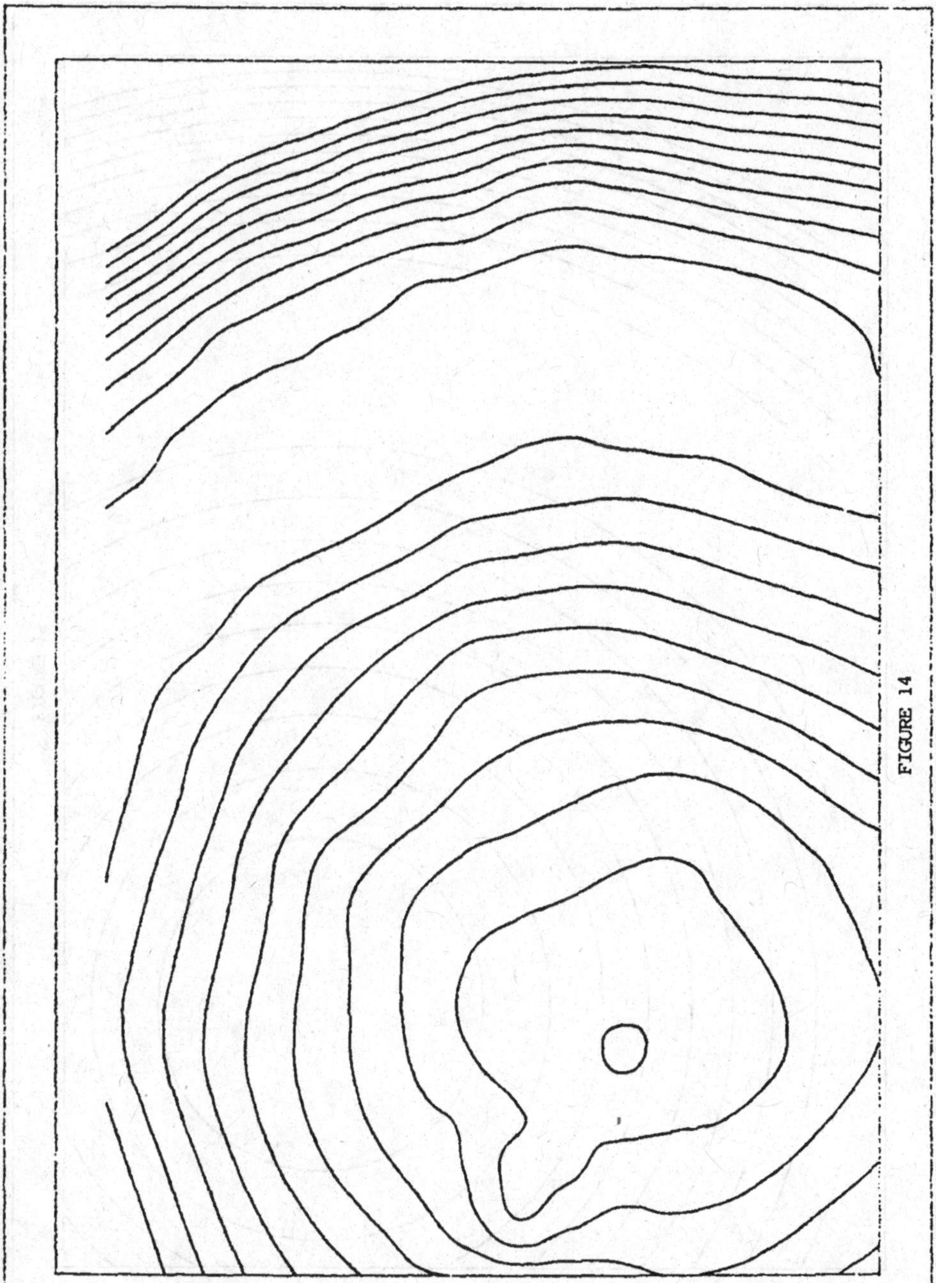

FIGURE 14

UTRE LE 31/03/79 A 01:03:20 DESSIN NO 002

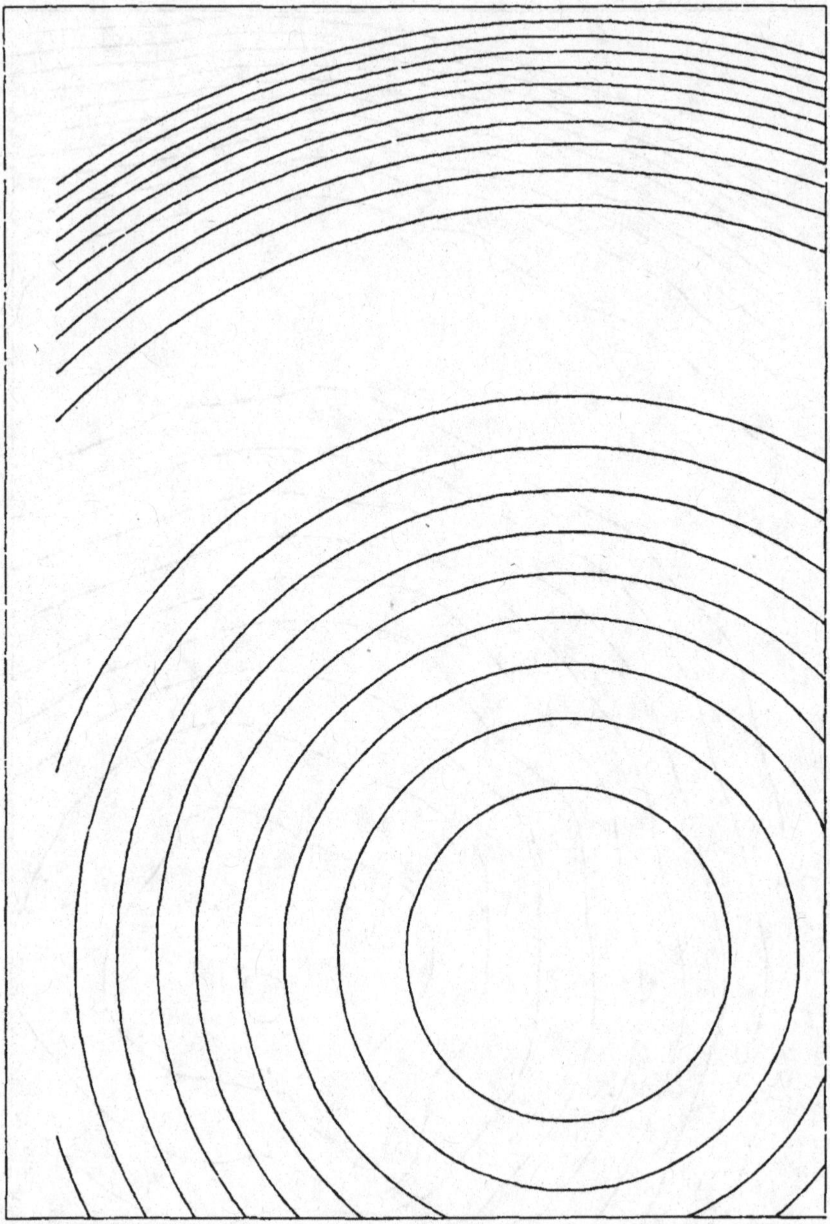

FIGURE 15

BIBLIOGRAPHIE :

[1] ATTEIA M. "Théorie et Applications de Fonctions Spline en Analyse Numérique" Thèse. Grenoble (1966)

[2] BABUSKA I & AZIZ A.K. "Foundations of the Finite Element Method with applications to Partial Differential Equations" Edited by A.K. AZIZ Academic Press, New York (1972)

[3] CHATELIN F. "Théorie de l'Approximation des Opérateurs Linéaires. Application au Calcul des Valeurs Propres d'Operateurs Differentiels et Integraux". Cours de DEA d'Analyse Numérique. Université Scientifique et Médicale de Grenoble (1966)

[4] COURANT R. & HILBERT D. "Methods of Mathematical Physics" Vol I. Interscience Publischers, New York (1953)

[5] CRAVEN P. & WAHBA G. "Smoothing Noisy Data with Spline Functions. Estimating the Correct Degree of Smoothing by the Method of Generalized Cross-Validation" Numerische Mathematik 31, 317-403 (1979)

[6] DUCHON J. "Fonctions Spline à Energie Invariante par Rotation" Rapport de Recherche n° 27 U.S.M.G. Grenoble (1976)

[7] DUCHON J. "Interpolation des Fonctions de Deux Variables par des Fonctions Spline du Type Plaque Mince" R.A.I.R.O. Analyse Numérique. Vol 10 - n° 12 (1976) pp 5-12

[8] FIX G. "Effects of Quadrature Errors in Finite Element Approximation of Steady State, Eigenvalue and Parabolic Problems" in " Foundations of the Finite Element Method with Applications to Partial Differential Equations" Edited by AZIZ A.K. Academic Press, New York (1972)

[9] GOLUB G., HEATH M. & WAHBA G. "Generalized Cross-Validation as a Method for Choosing a Good Ridge Parameter" Technical Report n° 491 May 1977 University of Winsconsin Madison.

[10] KREISS H.O. "Difference Approximations for Boundary and Eigenvalue Problems for Ordinary Differential Equations" Mathematics of Computations, Vol 26, 1972, pp 605-624

[11] LAURENT P.J. "Approximation et Optimisation" Hermman, Paris, 1972

[12] NECAS J. "Les Méthodes Directes en Théorie des Equations Elliptiques" Masson, Paris, 1967

[13] PAIHUA L. "Quelques Méthodes Numériques pour les Fonctions Spline à une et deux Variables" Thèse Grenoble. Mai 1978

[14] PAIHUA L. & UTRERAS F. "Un Ensemble de Programmes pour l'Interpolation de Fonctions, par des Fonctions Spline du Type Plaque Mince" Rapport de Recherche n° 140. Octobre 1978. IRMA Grenoble.

[15] SCHWARTZ L. "Théorie des Distributions" Hermman, Paris, 1966.

[16] SCHWARZ H.R. "Tridiagonalization of a Symmetric Band Matrix" in "Linear Algebra" Edited by J.H. WILKINSON et C. REINSCH Springer-Verlag. Berlin 1971

[17] THOMMAN J. "Determination et Construction de Fonctions Spline à Deux Variables Définies sur un Domaine Rectangulaire ou Circulaire" Thèse Lille 1970.

[18] UTRERAS F. "Sur le Choix du Paramètre d'Ajustement dans le Lissage par Fonctions Spline"
Séminaire d'Analyse Numérique n° 296 Grenoble Mars 1978.

[19] UTRERAS F. "Sur le Choix du Paramètre d'Ajustement dans le Lissage par Fonctions Spline"
Exposé au Colloque National d'Analyse Numérique de France.
Giens , le 21 Mai 1978 soumis pour publication.

[20] UTRERAS F. "Utilisation des Programmes de Calcul du Paramètre d'Ajustement dans le Lissage par Fonctions Spline"
Rapport de Recherche n° 121 - Grenoble Mai 1978.

[21] UTRERAS F. "Quelques Résultats d'Optimalité pour la Méthode de Validation Croisée"
Séminaire d'Analyse Numérique n° 301 Grenoble Novembre 1978

[22] UTRERAS F. "Utilisation de la Méthode de Validation Croisée pour le Lissage par Fonctions Spline à une ou deux Variables"
Doctoral Disertation. Université Scientifique et Médicale de Grenoble.
Grenoble Mai 1979.

[23] WAHBA G. "Practical Approximate Solutions to Linear Operator Equations when the Data are Noisy"
SIAM Journal on Numerical Analysis - Vol 14 - n° 4 Sept. 1977.

[24] WAHBA G. "Smoothing Noisy Data with Spline Functions"
Numerische Mathematik 24 (1975) pp 383-393

[25] WAHBA G. "Optimal Smoothing of Density Estimates"
in Classification and Clustering, J. VAN RYZIN ed., 423-458, Academic Press (1977)

[26] WAHBA G. & WOLD S. "A Completely Automatic French Curve : Fitting Spline Functions by Cross-Validation"
Comm. Statistics 4, 1-7, 1975.

CONVERGENCE RATES OF "THIN PLATE" SMOOTHING SPLINES WHEN THE DATA ARE NOISY

(Preliminary report)

Grace Wahba

Department of Statistics

University of Wisconsin

Madison, Wisconsin 53705 U.S.A.

Abstract

We study the use of "thin plate" smoothing splines for smoothing noisy d dimensional data. The model is

$$z_i = u(t_i)+\varepsilon_i \ , \qquad i = 1,2,\dots,n \ ,$$

where u is a real valued function on a closed, bounded subset Ω of Euclidean d-space and the ε_i are random variables satisfying $E\varepsilon_i=0$, $E\varepsilon_i\varepsilon_j=\sigma^2$, $i=j$, $=0$, $i\neq j$, $t_i\in\Omega$. The z_i are observed. It is desired to estimate u, given $z_1,\dots,z_n$. u is only assumed to be "smooth", more precisely we assume that u is in the Sobolev space $H^m(\Omega)$ of functions with partial derivatives up to order m in $L_2(\Omega)$, with $m>d/2$. u is estimated by $u_{n,m,\lambda}$, the restriction to Ω of $\tilde{u}_{n,m,\lambda}$, where $\tilde{u}_{n,m,\lambda}$ is the solution to: Find $\tilde{u}$ (in an appropriate space of functions on R^d) to minimize

$$\frac{1}{n}\sum_{i=1}^{n}(\tilde{u}(t_i)-z_i)+\lambda \sum_{i_1,\dots,i_m=1}^{d} \int_{R^d} \left(\frac{\partial^m \tilde{u}}{\partial x_{i_1}\partial x_{i_2}\dots\partial x_{i_m}}\right)^2 dx_1,dx_2,\dots,dx_d \ .$$

This minimization problem is known to have a solution for $\lambda>0$, $m>\frac{d}{2}$, $n\geq M=\binom{m+d-1}{d}$, provided the $t_1,\dots,t_n$ are "unisolvent". We consider the integrated mean square error

$$R(\lambda) = \frac{1}{|\Omega|}\int_\Omega (u_{n,m,\lambda}(t)-u(t))^2 dt \ , \qquad |\Omega| = \int_\Omega dt \ ,$$

and $ER(\lambda)$, as $\{t_i\}_{i=1}^n$ become dense in Ω. An estimate of λ which asymptotically minimizes $ER(\lambda)$ can be obtained by the method of generalized cross-validation. In this paper we give plausible arguments and numerical evidence supporting the following conjectures:

Suppose $u \in H^m(\Omega)$. Then

$$\min_\lambda ER(\lambda) = O(n^{-2m/(2m+d)}) \ .$$

Research supported by the Office of Naval Research under Grant No. N00014-77-C-0675.

Suppose $u \epsilon H^{2m}(\Omega)$ and certain other conditions are satisfied. Then -

$$\min_{\lambda} ER(\lambda) = O(n^{-4m/(4m+d)}) .$$

1. Introduction

Consider the model

$$z_i = u(t_i)+\varepsilon_i\ , \qquad i = 1,2,\ldots,n \tag{1.1}$$

where u is some "smooth" function on Ω, a closed, bounded subset of R^d, and the $\{\varepsilon_i\}$ are independent, zero mean random variables with common unknown variance σ^2. The $t_1,\ldots,t_n$ are in Ω, and $\underset{\sim}{z} = (z_1,\ldots,z_n)'$ is observed. It is desired to estimate u nonparametrically from $\underset{\sim}{z}$.

Our estimate $u_{n,m,\lambda}$ for u will be obtained as follows:
Let $\tilde{u}_{n,m,\lambda}$ be the solution the the following minimization problem: Find $\tilde{u} \in \underline{X}$ to minimize

$$\frac{1}{n}\sum_{i=1}^{n}(\tilde{u}(t_i)-z_i)^2+\lambda \sum_{i_1,\ldots,i_m=1}^{d} \int_{R_d} \left(\frac{\partial^m\tilde{u}}{\partial x_{i_1}\ldots\partial x_{i_m}}\right)^2 dx_1\ldots dx_d\ .$$

For example, when d=2, m=2, the second or "smoothness penalty" term becomes

$$\lambda \int_{R^2} (\tilde{u}^2_{x_1x_1}+2\tilde{u}^2_{x_1x_2}+\tilde{u}^2_{x_2x_2})dx_1dx_2\ ,$$

which is the bending energy of a thin plate. The space $\underline{X}$ is the "Beppo Levi" space

$$\underline{X} = \{\tilde{u}\varepsilon \mathcal{D}',\ \frac{\partial^{\alpha_1+\ldots+\alpha_d}\tilde{u}}{\partial x_1^{\alpha_1}\ldots\partial x_d^{\alpha_d}} \ \varepsilon\ L_2(R^d),\ \text{for } \alpha_1+\ldots+\alpha_d=m\}$$

where $\mathcal{D}'$ is the dual of the Schwartz space $\mathcal{D}$ of infinitely differentiable functions with compact support. See Meinguet (1978,1979) for further details. $u_{n,m,\lambda}$ is taken as the restriction of $\tilde{u}_{n,m,\lambda}$ to Ω.

A unique (continuous) solution is known to exist for any $\lambda>0$ provided

$$m > d/2$$
$$n\geq M = \binom{m+d-1}{d}$$

and the "design" $t_1,\ldots,t_n$ is "unisolvent", that is, if $\{\phi_\nu\}_{\nu=1}^M$ are a basis for the M dimensional space of polynomials of total degree m-1 or less, then $\sum_{\nu=1}^{M} a_\nu\phi_\nu(t_i) = 0$, $i = 1,2,\ldots,n$, implies that the a_ν are all 0. See Duchon (1976a,1976b), Meinguet (1978,1979), Paihua (1977,1978). We henceforth assume these conditions. Duchon has shown that the solution has a representation

$$\tilde{u}_{n,m,\lambda}(t) = \sum_{j=1}^{n} c_jE_m(t,t_j) + \sum_{\nu=1}^{M} d_\nu\phi_\nu(t)\ ,$$

where

$$E_m(s,t) = \theta_m |s-t|^{2m-d} \log|s-t| \qquad m \text{ even}$$

$$= \theta_m |s-t|^{2m-d} \qquad m \text{ odd}$$

where, if $s = (x_1,\ldots,x_d)$, $t = (y_1,\ldots,y_d)$, $|s-t| = (\sum_{i=1}^{d}(x_i-y_i)^2)^{1/2}$, and

$$\theta_m = (-1)^{d/2+1}/(2^{2m-1}\pi^{d/2}(m-1)!\,(m-d/2)!) \qquad m \text{ even}$$

$$= (-1)^m \Gamma(d/2-m)/2^{2m}\pi^{d/2}(m-1)! \qquad m \text{ odd}.$$

The coefficients $\underset{\sim}{c} = (c_1,\ldots,c_n)'$ and $\underset{\sim}{d} = (d_1,\ldots,d_M)'$ are determined by

$$(K+\rho I)\underset{\sim}{c}+T\underset{\sim}{d} = \underset{\sim}{z} \tag{1.2}$$

$$T'\underset{\sim}{c} = 0, \tag{1.3}$$

where K is the $n\times n$ matrix with jk^{th} entry $E_m(t_j,t_k)$, $\rho=n\lambda$, T is the $n\times M$ matrix with $i\nu^{th}$ entry $\phi_\nu(t_i)$ and $\underset{\sim}{z} = (z_1,\ldots,z_n)'$. See Duchon (1976,1977), Paihua (1977,1978), Wahba (1979). We discuss the choice of λ shortly.

Let Ω be a closed, bounded subset in R^d. We will suppose that the $\{t_i\}$ become dense in Ω in such a way that

$$\lim \frac{1}{n}\sum_{i=1}^{n} \rho(t_i) = \frac{1}{|\Omega|}\int_\Omega \rho(t)dt, \qquad |\Omega| = \int_\Omega dt \tag{1.4}$$

for any continuous ρ. (However, it will be clear that our rate arguments hold under weaker conditions on the distribution of the $\{t_i\}$, for example

$$\lim \frac{1}{n}\sum_{i=1}^{n} \rho(t_i) = \int_\Omega \rho(t)w(t)dt$$

for some sufficiently nice positive w.) Let $R(\lambda)$ be the integrated mean square erro when λ is used:

$$R(\lambda) = \frac{1}{n}\sum_{i=1}^{n}(u_{n,m,\lambda}(t_i)-u(t_i))^2 \approx \frac{1}{|\Omega|}\int_\Omega (u_{n,m,\lambda}(t)-u(t))^2 dt. \tag{1.5}$$

The smoothing parameter λ^* which minimizes $ER(\lambda)$ can be estimated by the method of generalized cross-validation (GCV), see Craven and Wahba (1979), Golub, Heath and Wahba (1977), Wahba (1979). Pleasing numerical results have been obtained in Monte Carlo studies for d=1, m=2 (Craven and Wahba (1979)) and d=2, m=2, Wahba (1979). Convergence rates for $ER(\lambda^*)$ have been obtained in the one dimensional case (Wahba (1975)).

Stone (1978) has recently obtained some rather general results on best achieveable pointwise convergence rates for the model (1.1), for any method of estimation of u(t). Reduced to our case and phrased loosely, his results say that the rate

$$E(\hat{u}(t)-u(t))^2 = O(n^{-(2m-1)/(2m-1+d)}) ,$$

where $\hat{u}(t)$ is any estimate of $u(t)$ from the data $\underset{\sim}{z}$, can be achieved for all $u \in H_m(\Omega)$ but not bettered. In this paper we are concerned with integrated mean square error convergence rates: —

$$E \frac{1}{|\Omega|} \int_\Omega (u_{n,m,\lambda^*}(t)-u(t))^2 dt = ER(\lambda^*)$$

of u_{n,m,λ^*} .

It is our goal to give a plausible argument that

i) if $u \in H^m(\Omega)$,

$$ER(\lambda^*) = O(n^{-2m/(2m+d)})$$

and if

ii) $u \in H^{2m}(\Omega)$ and some other conditions are satisfied, then

$$ER(\lambda^*) = O(n^{-4m/(4m+d)})$$

Our argument follows the arguments given in Wahba (1975,1977) and Craven and Wahba (1979), and is given in section 2.

2. Plausibility arguments, numerical evidence

Let $A(\lambda)$ be the $n\times n$ matrix defined by

$$\begin{pmatrix} u_{n,m,\lambda}(t_1) \\ \vdots \\ u_{n,m,\lambda}(t_n) \end{pmatrix} = A(\lambda)\underset{\sim}{z} .$$

If $R(\lambda)$ is taken as the middle quantity in (1.5), we have

$$R(\lambda) = \frac{1}{n} ||A(\lambda)(\underset{\sim}{u}+\underset{\sim}{\varepsilon})-\underset{\sim}{u}||^2$$

where $\underset{\sim}{u} = (u(t_1),\ldots,u(t_n))'$, $\underset{\sim}{\varepsilon} = (\varepsilon_1,\ldots,\varepsilon_n)'$, and

$$ER(\lambda) = \frac{1}{n} ||(I-A(\lambda))\underset{\sim}{u}||^2 + \frac{\sigma^2}{n} \text{Trace } A^2(\lambda) . \qquad (2.1$$

($A(\lambda)$ is symmetric.)

We call $\frac{1}{n} ||(I-A(\lambda))\underset{\sim}{u}||^2$ the "squared bias" and (σ^2/n) Trace $A^2(\lambda)$ the variance

<u>Lemma 1</u>.

$$\frac{1}{n} ||(I-A(\lambda))\underset{\sim}{u}||^2 \le \lambda J_m(\tilde{u}) \qquad (2.2$$

where, for $\tilde{v} \in \underline{X}$

$$J_m(\tilde{v}) = \sum_{i_1,\ldots,i_m=1}^{d} \int_{R^d} \left(\frac{\partial^m \tilde{v}(x_1,\ldots,x_d)}{\partial x_{i_1}\ldots\partial x_{i_m}}\right)^2 dx_1\ldots dx_d$$

and $\tilde{u}$ is that element in $\underline{X}$ which minimizes J_m subject to coinciding with u on Ω.

<u>Proof</u>.

For each i, $\tilde{u}(t_i) = u(t_i)$. $A(\lambda)\underset{\sim}{u}$ is a vector of values of the function, call it $\tilde{u}^*_{n,m,\lambda}$ which is the solution to the problem: Find $\tilde{v} \in \underline{X}$ to minimize

$$\frac{1}{n} \sum_{j=1}^{n} (u(t_j)-\tilde{v}(t_j))^2 + \lambda J_m(\tilde{v}) .$$

Therefore

$$\frac{1}{n} \sum_{j=1}^{n} (u(t_j)-\tilde{u}^*_{n,m,\lambda}(t_j))^2 + \lambda J_m(\tilde{u}^*_{n,m,\lambda})$$

$$= \frac{1}{n} ||(I-A)\underset{\sim}{u}||^2 + \lambda J_m(\tilde{u}^*_{n,m,\lambda})$$

$$\le \frac{1}{n} \sum_{j=1}^{n} (u(t_j)-\tilde{u}(t_j))^2 + \lambda J_m(\tilde{u}) = \lambda J_m(\tilde{u}) .$$

We now investigate Trace $A^2(\lambda)$. Let $T_{n\times M}$ be the $n\times M$ dimensional matrix with $j\nu^{th}$ entry $\phi_\nu(t_j)$. Let R by <u>any</u> $n\times(n-M)$ dimensional matrix of rank n-M satisfying $R'T = 0_{(n-M)\times M}$, $R'R = I_{n-M}$. Following the results of Anselone and Laurent (1968) it is shown in Wahba (1979) that $\underset{\sim}{c}$ and $\underset{\sim}{d}$ satisfying (1.2) and (1.3) have the representations

$$\underset{\sim}{c} = R(R'KR+\rho I)^{-1}R'\underset{\sim}{z}$$

$$\underset{\sim}{d} = (T'T)^{-1}T'(\underset{\sim}{z}-K\underset{\sim}{c})$$

and that

$$(I-A(\lambda))\underset{\sim}{z} = \rho\underset{\sim}{c} = n\lambda R(R'KR+n\lambda I)^{-1}R'\underset{\sim}{z} \;, \qquad \underset{\sim}{z} \in E_n \;. \tag{2.3}$$

Hence, if we define $B = R'KR$ and let $b_{\nu n}$, $\nu = 1,2,\ldots,n-M$ be the m-M eigenvalues of B, then

$$\frac{1}{n}\,\mathrm{Tr}\,A^2(\lambda) = \frac{1}{n}\sum_{\nu=1}^{n}\left(\frac{b_{\nu n}}{b_{\nu n}+n\lambda}\right)^2 = \frac{1}{n}\sum_{\nu=1}^{n}\frac{1}{(1+n\lambda/b_{\nu n})^2} \;.$$

We remark that K is not, in general, positive definite, however R'KR is, since it is known that $\underset{\sim}{r}'K\underset{\sim}{r}>0$ for any non-trivial $\underset{\sim}{r}$ satisfying $T'\underset{\sim}{r} = 0$ (See Paihua (1977), Duchon (1977)).

<u>Lemma 2</u>.

Suppose there exist $p>1$, and k_1, k_2 with $0 < k_1 \le k_2 < \infty$ such that

$$\frac{k_1}{\nu^p} \le \frac{b_{\nu n}}{n} \le \frac{k_2}{\nu^p}$$

then, for some constant k_3,

$$\frac{1}{n}\,\mathrm{Tr}\,A^2(\lambda) = \frac{k_3}{n\lambda^{1/p}}\,(1+o(1)) \;. \tag{2.4}$$

<u>Outline of Proof</u>.

$$\frac{1}{n}\,\mathrm{Tr}\,A^2(\lambda) = \frac{1}{n}\sum_{\nu=1}^{n}\frac{1}{(1+\lambda n/b_{\nu n})} = \frac{1}{n}\sum_{\nu=1}^{n}\frac{1}{(1+k_3^{-1}\lambda\nu^p)^2}\,(1+o(1))$$

$$= \frac{1}{n}\int_0^\infty \frac{dx}{(1+k_3^{-1}\lambda x^p)^2}\,(1+o(1))$$

$$= \frac{k_3}{n\lambda^{1/p}}\,(1+o(1)) \quad \text{for some } k_3 \in [k_1,k_2] \;.$$

(A more rigorous argument can be found in Craven and Wahba 1979)).

Lemma 3. (Conjecture)

For $2m/d>1$ there exist k_1, k_2 with $0 < k_1 \le k_2 < \infty$ such that

$$\frac{k_1}{\nu^{2m/d}} \le \frac{b_{\nu n}}{n} \le \frac{k_2}{\nu^{2m/d}} \tag{2.5}$$

Argument

We first argue that the eigenvalues $\lambda_1, \lambda_2, \ldots$ of the integral operator K on $L_2(\Omega)$ defined by

$$(Ku)(t) = \int_\Omega E_m(t,s)u(s)ds$$

go to 0 at the rate $\nu^{-2m/d}$, and then that this entails that the eigenvalues $b_{\nu n}$ of K behave like $n\lambda_\nu/|\Omega|$, $\nu = 1,2,\ldots,n$, $n = 1,2,\ldots$.

Δ^m is a left inverse of K, since, if

$$\psi(t) = \int_\Omega E_m(t,s)\phi(s)ds$$

then $\Delta^m\psi(t) = \phi(t)$, $t\varepsilon\Omega$ (See Courant and Hilbert (1953)). Thus it is to be expected that the eigenvalues of K asymptotically decrease at the same rate as the eigenvalues of Δ^m increase. Let $d = 2$ and suppose Ω is the rectangle with sides a_1 and a_2. The eigenfunctions $\{\phi\}$ and eigenvalues $\{\rho\}$ for the equation

$$\Delta u = \rho u$$

with boundary conditions $u = 0$ on $\partial\Omega$ are

$$\phi_{\xi,\eta}(x_1,x_2) = \sin\frac{\xi\pi x_1}{a_1}\sin\frac{\eta\pi x_2}{a_2}$$

$$\rho_{\xi\eta} = \pi^2\left(\frac{\xi^2}{a_1^2} + \frac{\eta^2}{a_2^2}\right), \qquad \xi,\eta = 1,2,\ldots$$

It follows, by counting the number of pairs (ξ,η) in the ellipse $\pi^2\left(\frac{x_1^2}{a_1^2} + \frac{x_2^2}{a_2^2}\right) \le c$, that, if the eigenvalues $\rho_{\xi\eta}$ $(\xi,\eta = 1,2,\ldots)$ are reindexed in size place as ρ_ν, $\nu = 1,2,\ldots$ that

$$\lim_{\nu\to\infty}\frac{\rho_\nu}{\nu} = \frac{4\pi}{|\Omega|} .$$

This relation is known to hold independently of the shape of Ω, and also for a Neumann boundary condition instead of $u = 0$ on $\partial\Omega$. Similarly eigenfunctions and eigenvalues for

$$\Delta^m u = \rho u$$

$$\Delta^k u = 0 \text{ on } \partial\Omega,\ k = 0,1,\ldots,m-1$$

are $\phi_{\xi,\eta}$ and $\rho^m_{\xi\eta}$ so that the eigenvalues $\{\rho_\nu\}$ satisfy

$$\lim_{\nu\to\infty} \frac{\rho_\nu}{\nu^m} = \left(\frac{4\pi}{|\Omega|}\right)^m ,$$

and this result is independent of the shape of Ω. Going to d = 3 dimensions, the eigenvalues for $\Delta u = \rho u$ on a rectangle with sides a_1, a_2 and a_3 and suitable boundary conditions are

$$\rho_{\xi,\eta,\psi} = \pi^2 \left(\frac{\xi^2}{a_1^2} + \frac{\eta^2}{a_2^2} + \frac{\psi^2}{a_3^2}\right) \qquad \xi,\eta,\psi = 1,2,\ldots$$

and, by counting the number of triplets (ξ,η,ψ) in the ellipse

$$\pi^2 \left(\frac{x_1^2}{a_1^2} + \frac{x_2^2}{a_2^2} + \frac{x_3^2}{a_3^2}\right) \le c$$

one obtains that

$$\lim_{\nu\to\infty} \frac{\rho_\nu^{3/2}}{\nu} = \frac{6\pi^2}{|\Omega|} .$$

or

$$\frac{\rho_\nu}{\nu^{2/3}} = \left(\frac{6\pi^2}{|\Omega|}\right)^{2/3}(1+o(1)) .$$

See Courant and Hilbert (1953). Similarly the eigenvalues for Δ^m satisfy

$$\frac{\rho_\nu}{\nu^{2m/3}} = \left(\frac{6\pi^2}{|\Omega|}\right)^{2m/3}(1+o(1))$$

and, extending the argument to d dimensions gives

$$\frac{\rho_\nu}{\nu^{2m/d}} = \left(\frac{(2\pi)^d}{V_d|\Omega|}\right)^{2m/d}(1+o(1))$$

where V_d is the volume of the sphere of radius 1 in d dimensions. Therefore, we conjecture that the rate of decrease of the eigenvalues $\{\lambda_\nu\}$ of K is $\nu^{-2m/d}$.

Let K(s,t) be a kernel with a Mercer-Hilbert Schmidt expansion on Ω,

$$K(s,t) = \sum_{\nu=1}^{\infty} \lambda_\nu \phi_\nu(s)\phi_\nu(t)$$

where the eigenvalues $\{\lambda_\nu\}$ are absolutely summable and the eigenfunctions $\{\phi_\nu\}$ are an orthonormal set on $L_2(\Omega)$. Then, for large n,

$$K(t_i,t_j) \approx \sum_{\nu=1}^{n} n\lambda_\nu (\phi_\nu(t_i)/\sqrt{n})(\phi_\nu(t_j)/\sqrt{n}) ,$$

and provided

$$\frac{|\Omega|}{n} \sum_{i=1}^{n} \phi_\nu(t_i)\phi_\mu(t_i) \simeq \int_\Omega \phi_\nu(t)\phi_\mu(t)dt = 1, \ \mu = \nu$$
$$= 0, \ \mu \neq \nu ,$$

we see that the eigenvalues $\lambda_{\nu n}$, $\nu = 1,2,\ldots,n$, say, of the matrix K with jk^{th} entry

$K(t_j,t_k)$, have an approximation as $\lambda_{\nu n} \sim n\lambda_\nu/|\Omega|$.

We have computed the eigenvalues $b_{\nu n}$, $\nu = 1,2,\ldots,n-M$ for an example with $d = 2$, $m = 2$, and $n = 81$. Thus there are $n-M = 78$ eigenvalues. The t_i are arranged on a 9×9 square array. If $b_{\nu n} \sim c\nu^{-p}$, then a plot of $b_{\nu n}$ vs. ν on log-log paper should fall on a straight line with slope $-p$, here $p = 2$. Figure 1 gives a plot of these 78 eigenvalues. For comparison, a solid line has been drawn with slope -2.

Theorem

Suppose Lemma 3 is true. Then, if $u \in H^m(\Omega)$,

$$\min_\lambda ER(\lambda) = O(n^{-2m/(2m+d)}) .$$

Proof

By (2.1), (2.2), (2.4) and (2.5)

$$ER(\lambda) \le c_1\lambda + c_2/n\lambda^{d/2m}$$

where c_1 and c_2 are constants. Minimizing this expression with respect to λ gives $\lambda^* = O(n^{-2m/(2m+d)})$, where λ^* is the minimizer of $R(\lambda)$, and thence the result.

Lemma 4. (Conjecture)

Suppose u has a representation

$$u(t) = \int_\Omega E_m(t,s)\rho(s)ds + \sum_{\nu=1}^{M} \theta_\nu\phi_\nu(t) \tag{2.6}$$

where ρ is piecewise continuous and satisfies $\int_\Omega \phi_\nu(s)\rho(s)ds = 0$, $\nu = 1,2,\ldots,M$. Then

$$\frac{1}{n}\,||(I-A(\lambda))\underset{\sim}{u}||^2 \le \lambda^2\,|\Omega|\,\int_\Omega (\Delta^m u)^2 dt(1+o(1)) . \tag{2.7}$$

Remark: If u has the given form, then $\rho = \Delta^m u$. However, the class of functions with representation (2.6) is restrictive since, for example it excludes harmonic functions other than polynomials of degree $\le m$. (See Courant and Hilbert)

Suggestion of proof. From (2.3) we have

$$\frac{1}{n}\,||(I-A(\lambda))\underset{\sim}{u}||^2 = n\lambda^2\underset{\sim}{u}'R(RKR'+n\lambda I)^{-2}R'\underset{\sim}{u}$$

and the right hand side is bounded above by

$$n\lambda^2\,||(RKR')^{-1}R'\underset{\sim}{u}||^2$$

If u has the required form (2.6), then

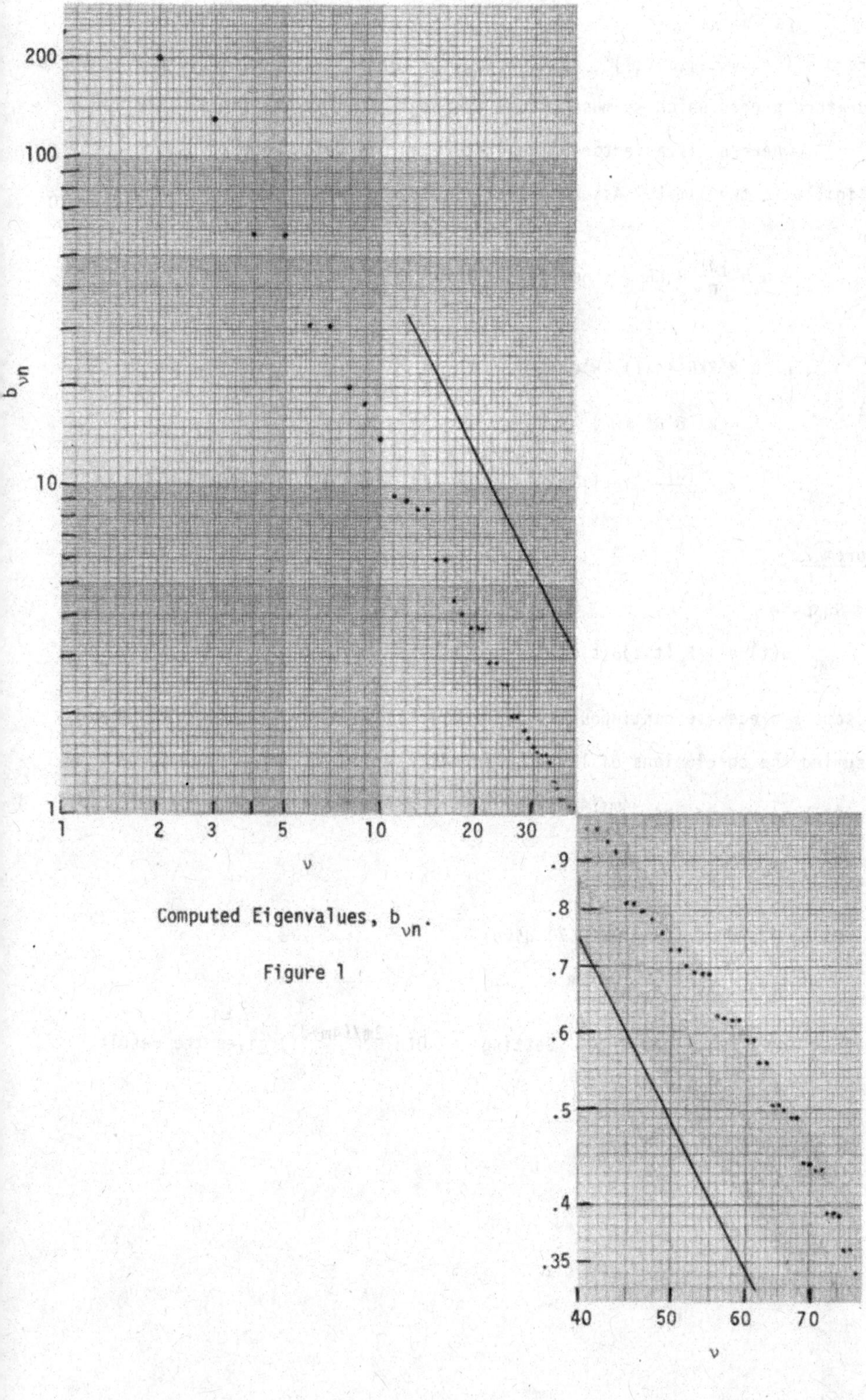

Computed Eigenvalues, $b_{\nu n}$.

Figure 1

$$u = \frac{|\Omega|}{n} K\rho + T\theta + \delta^1$$

where $\rho = (\rho(t_1),\ldots,\rho(t_n))'$, $\theta = (\theta_1,\ldots,\theta_M)'$ and $\delta' = (\delta_1^1,\ldots,\delta_n^1)$ is a vector of quadrature errors which we must assume are negligible in the limit. Similarly $T'\rho = \delta^2$, where δ^2 is a vector of quadrature errors which we must assume are negligible in the limit. Assuming $T'\rho = 0$, then $\rho = R\rho_0$, for some n-M vector ρ_0. Then

$$R'u = \frac{|\Omega|}{n} R'KR\rho_0 + \text{negligible terms}$$

and

$$n\lambda^2 u'R(RKR'+n\lambda I)^{-2}R'u$$

$$\leq n\lambda^2 u'R(RKR')^{-2}R'u$$

$$= \frac{\lambda^2|\Omega|^2}{n} ||\rho_0||^2 = \frac{\lambda^2|\Omega|^2}{n} ||\rho||^2 = \lambda^2|\Omega| \int_\Omega \rho^2(t)dt(1+o(1)) .$$

Theorem 2.

Suppose

$$u(t) = \int_\Omega E_m(t,s)\rho(s)ds + \sum_{\nu=1}^{M} \theta_\nu\phi_\nu(t)$$

for some ρ piecewise continuous with $\int_\Omega \phi_\nu(s)\rho(s)ds = 0$, $\nu = 1,2,\ldots,M$. Then (assuming the conclusions of lemmas 3 and 4),

$$\min_\lambda R(\lambda) = O(n^{-4m/(4m+d)})$$

Proof.

Using (2.4), (2.5) and (2.7) gives

$$R(\lambda) \leq k_3\lambda^2 + k_4/n\lambda^{d/2m}$$

where k_3 and k_4 are constants. Setting $\lambda = O(n^{-2m/(4m+d)})$ gives the result.

References

Anselone, P.M., and Laurent, P.J. (1968). A general method for the construction of interpolating or smoothing spline-functions, Numer. Math. 12, 66-82.

Aubin, Jean-Pierre (1972). Approximation of Elliptic Boundary Value Problems, Wiley-Interscience.

Courant, R. and Hilbert, D. (1953). Methods of Mathematical Physics, Vol. I, Interscience.

Craven, P., and Wahba, G. (1979). Smoothing noisy data with spline functions: estimating the correct degree of smoothing by the method of generalized cross-validation, Numer. Math. 31, 377-403.

Duchon, Jean (1976a). Interpolation des fonctions de deux variables suivant le principe de la flexion des plaques minces. R.A.I.R.O. Analyse Numerique, 10, 12, pp.5-12.

Duchon, Jean (1976b). Fonctions spline du type "Plaque mince" en dimension 2. No. 231. Seminaire d'analyse numerique, Mathematiques Appliqueés, Université Scientifique et Medicale de Grenoble.

Duchon, Jean, (1977). Fonctions-spline a energie invariante par rotation, R.R.#27, Mathematiques Appliqueés, Université Scientifique et Medicale de Grenoble.

Golub, G., Heath, M., and Wahba, G. (1977). Generalized cross-validation as a method for choosing a good ridge parameter, University of Wisconsin-Madison, Department of Statistics Technical Report #491, to appear, Technometrics.

Meinguet, Jean, (1978). Multivariate interpolation at arbitrary points made simple. Report No. 118, Institute de Mathematique Pure et Appliqueé, Université Catholique de Louvain, to appear, ZAMP.

Meinguet, Jean, (1979). An intrinsic approach to multivariate spline interpolation at arbitrary points. To appear in the Proceedings of the NATO Advanced Study Institute on Polynomial and Spline Approximation, Calgary 1978, B. Sahney, Ed.

Paihua, Montes, L. (1977). Methodes numeriques pour l'obtention de fonctions-spline du type plaque mince en dimension 2. Seminaire d'Analyse Numerique No. 273, Mathematiques Appliqueés, Université Scientifique et Medicale de Grenoble.

Paihua Montes, Luis, (1978). Quelques methodes numeriques pour le calcul de fonctions splines a une et plusiers variables. Thesis, Analyse Numerique Université Scientifique et Medicale de Grenoble.

Stone, Charles (1978). Optimal rates of convergence for non-parametric estimators (manuscript).

Wahba, G. (1975). Smoothing noisy data with spline functions, Numer. Math., 24, 383-393.

Wahba, G. (1977). Discussion to C. J. Stone Consistent nonparametric regression, Ann. Statist. 5, 637-640.

Wahba, Grace (1979). How to smooth curves and surfaces with splines and cross-validation. University of Wisconsin-Madison, Department of Statistics TR#555.

Vol. 580: C. Castaing and M. Valadier, Convex Analysis and Measurable Multifunctions. VIII, 278 pages. 1977.

Vol. 581: Séminaire de Probabilités XI, Université de Strasbourg. Proceedings 1975/1976. Edité par C. Dellacherie, P. A. Meyer et M. Weil. VI, 574 pages. 1977.

Vol. 582: J. M. G. Fell, Induced Representations and Banach *-Algebraic Bundles. IV, 349 pages. 1977.

Vol. 583: W. Hirsch, C. C. Pugh and M. Shub, Invariant Manifolds. IV, 149 pages. 1977.

Vol. 584: C. Brezinski, Accélération de la Convergence en Analyse Numérique. IV, 313 pages. 1977.

Vol. 585: T. A. Springer, Invariant Theory. VI, 112 pages. 1977.

Vol. 586: Séminaire d'Algèbre Paul Dubreil, Paris 1975–1976 (29ème Année). Edited by M. P. Malliavin. VI, 188 pages. 1977.

Vol. 587: Non-Commutative Harmonic Analysis. Proceedings 1976. Edited by J. Carmona and M. Vergne. IV, 240 pages. 1977.

Vol. 588: P. Molino, Théorie des G-Structures: Le Problème d'Equivalence. VI, 163 pages. 1977.

Vol. 589: Cohomologie l-adique et Fonctions L. Séminaire de Géométrie Algébrique du Bois-Marie 1965–66, SGA 5. Edité par L. Illusie. XII, 484 pages. 1977.

Vol. 590: H. Matsumoto, Analyse Harmonique dans les Systèmes de Tits Bornologiques de Type Affine. IV, 219 pages. 1977.

Vol. 591: G. A. Anderson, Surgery with Coefficients. VIII, 157 pages. 1977.

Vol. 592: D. Voigt, Induzierte Darstellungen in der Theorie der endlichen, algebraischen Gruppen. V, 413 Seiten. 1977.

Vol. 593: K. Barbey and H. König, Abstract Analytic Function Theory and Hardy Algebras. VIII, 260 pages. 1977.

Vol. 594: Singular Perturbations and Boundary Layer Theory, Lyon 1976. Edited by C. M. Brauner, B. Gay, and J. Mathieu. VIII, 539 pages. 1977.

Vol. 595: W. Hazod, Stetige Faltungshalbgruppen von Wahrscheinlichkeitsmaßen und erzeugende Distributionen. XIII, 157 Seiten. 1977.

Vol. 596: K. Deimling, Ordinary Differential Equations in Banach Spaces. VI, 137 pages. 1977.

Vol. 597: Geometry and Topology, Rio de Janeiro, July 1976. Proceedings. Edited by J. Palis and M. do Carmo. VI, 866 pages. 1977.

Vol. 598: J. Hoffmann-Jørgensen, T. M. Liggett et J. Neveu, Ecole d'Eté de Probabilités de Saint-Flour VI – 1976. Edité par P.-L. Hennequin. XII, 447 pages. 1977.

Vol. 599: Complex Analysis, Kentucky 1976. Proceedings. Edited by J. D. Buckholtz and T. J. Suffridge. X, 159 pages. 1977.

Vol. 600: W. Stoll, Value Distribution on Parabolic Spaces. VIII, 216 pages. 1977.

Vol. 601: Modular Functions of one Variable V, Bonn 1976. Proceedings. Edited by J.-P. Serre and D. B. Zagier. VI, 294 pages. 1977.

Vol. 602: J. P. Brezin, Harmonic Analysis on Compact Solvmanifolds. VIII, 179 pages. 1977.

Vol. 603: B. Moishezon, Complex Surfaces and Connected Sums of Complex Projective Planes. IV, 234 pages. 1977.

Vol. 604: Banach Spaces of Analytic Functions, Kent, Ohio 1976. Proceedings. Edited by J. Baker, C. Cleaver and Joseph Diestel. VI, 141 pages. 1977.

Vol. 605: Sario et al., Classification Theory of Riemannian Manifolds. XX, 498 pages. 1977.

Vol. 606: Mathematical Aspects of Finite Element Methods. Proceedings 1975. Edited by I. Galligani and E. Magenes. VI, 362 pages. 1977.

Vol. 607: M. Métivier, Reelle und Vektorwertige Quasimartingale und die Theorie der Stochastischen Integration. X, 310 Seiten. 1977.

Vol. 608: Bigard et al., Groupes et Anneaux Réticulés. XIV, 334 pages. 1977.

Vol. 609: General Topology and Its Relations to Modern Analysis and Algebra IV. Proceedings 1976. Edited by J. Novák. XVIII, 225 pages. 1977.

Vol. 610: G. Jensen, Higher Order Contact of Submanifolds of Homogeneous Spaces. XII, 154 pages. 1977.

Vol. 611: M. Makkai and G. E. Reyes, First Order Categorical Logic. VIII, 301 pages. 1977.

Vol. 612: E. M. Kleinberg, Infinitary Combinatorics and the Axiom of Determinateness. VIII, 150 pages. 1977.

Vol. 613: E. Behrends et al., L^P-Structure in Real Banach Spaces. X, 108 pages. 1977.

Vol. 614: H. Yanagihara, Theory of Hopf Algebras Attached to Group Schemes. VIII, 308 pages. 1977.

Vol. 615: Turbulence Seminar, Proceedings 1976/77. Edited by P. Bernard and T. Ratiu. VI, 155 pages. 1977.

Vol. 616: Abelian Group Theory, 2nd New Mexico State University Conference, 1976. Proceedings. Edited by D. Arnold, R. Hunter and E. Walker. X, 423 pages. 1977.

Vol. 617: K. J. Devlin, The Axiom of Constructibility: A Guide for the Mathematician. VIII, 96 pages. 1977.

Vol. 618: I. I. Hirschman, Jr. and D. E. Hughes, Extreme Eigen Values of Toeplitz Operators. VI, 145 pages. 1977.

Vol. 619: Set Theory and Hierarchy Theory V, Bierutowice 1976. Edited by A. Lachlan, M. Srebrny, and A. Zarach. VIII, 358 pages. 1977.

Vol. 620: H. Popp, Moduli Theory and Classification Theory of Algebraic Varieties. VIII, 189 pages. 1977.

Vol. 621: Kauffman et al., The Deficiency Index Problem. VI, 112 pages. 1977.

Vol. 622: Combinatorial Mathematics V, Melbourne 1976. Proceedings. Edited by C. Little. VIII, 213 pages. 1977.

Vol. 623: I. Erdelyi and R. Lange, Spectral Decompositions on Banach Spaces. VIII, 122 pages. 1977.

Vol. 624: Y. Guivarc'h et al., Marches Aléatoires sur les Groupes de Lie. VIII, 292 pages. 1977.

Vol. 625: J. P. Alexander et al., Odd Order Group Actions and Witt Classification of Innerproducts. IV, 202 pages. 1977.

Vol. 626: Number Theory Day, New York 1976. Proceedings. Edited by M. B. Nathanson. VI, 241 pages. 1977.

Vol. 627: Modular Functions of One Variable VI, Bonn 1976. Proceedings. Edited by J.-P. Serre and D. B. Zagier. VI, 339 pages. 1977.

Vol. 628: H. J. Baues, Obstruction Theory on the Homotopy Classification of Maps. XII, 387 pages. 1977.

Vol. 629: W. A. Coppel, Dichotomies in Stability Theory. VI, 98 pages. 1978.

Vol. 630: Numerical Analysis, Proceedings, Biennial Conference, Dundee 1977. Edited by G. A. Watson. XII, 199 pages. 1978.

Vol. 631: Numerical Treatment of Differential Equations. Proceedings 1976. Edited by R. Bulirsch, R. D. Grigorieff, and J. Schröder. X, 219 pages. 1978.

Vol. 632: J.-F. Boutot, Schéma de Picard Local. X, 165 pages. 1978.

Vol. 633: N. R. Coleff and M. E. Herrera, Les Courants Résiduels Associés à une Forme Méromorphe. X, 211 pages. 1978.

Vol. 634: H. Kurke et al., Die Approximationseigenschaft lokaler Ringe. IV, 204 Seiten. 1978.

Vol. 635: T. Y. Lam, Serre's Conjecture. XVI, 227 pages. 1978.

Vol. 636: Journées de Statistique des Processus Stochastiques, Grenoble 1977, Proceedings. Edité par Didier Dacunha-Castelle et Bernard Van Cutsem. VII, 202 pages. 1978.

Vol. 637: W. B. Jurkat, Meromorphe Differentialgleichungen. VII, 194 Seiten. 1978.

Vol. 638: P. Shanahan, The Atiyah-Singer Index Theorem, An Introduction. V, 224 pages. 1978.

Vol. 639: N. Adasch et al., Topological Vector Spaces. V, 125 pages. 1978.

Vol. 640: J. L. Dupont, Curvature and Characteristic Classes. X, 175 pages. 1978.

Vol. 641: Séminaire d'Algèbre Paul Dubreil, Proceedings Paris 1976–1977. Edité par M. P Malliavin. IV, 367 pages. 1978

Vol. 642: Theory and Applications of Graphs, Proceedings, Michigan 1976. Edited by Y. Alavi and D. R. Lick. XIV, 635 pages. 1978.

Vol. 643: M. Davis, Multiaxial Actions on Manifolds. VI, 141 pages. 1978.

Vol. 644: Vector Space Measures and Applications I, Proceedings 1977. Edited by R. M. Aron and S. Dineen. VIII, 451 pages. 1978.

Vol. 645: Vector Space Measures and Applications II, Proceedings 1977. Edited by R. M. Aron and S. Dineen. VIII, 218 pages. 1978.

Vol. 646: O. Tammi, Extremum Problems for Bounded Univalent Functions. VIII, 313 pages. 1978

Vol. 647: L. J. Ratliff, Jr., Chain Conjectures in Ring Theory. VIII, 133 pages. 1978.

Vol. 648: Nonlinear Partial Differential Equations and Applications, Proceedings, Indiana 1976–1977. Edited by J. M. Chadam. VI, 206 pages. 1978.

Vol. 649: Séminaire de Probabilités XII, Proceedings, Strasbourg, 1976–1977. Edité par C. Dellacherie, P. A. Meyer et M. Weil. VIII, 805 pages. 1978.

Vol. 650: C*-Algebras and Applications to Physics. Proceedings 1977. Edited by H. Araki and R. V. Kadison. V, 192 pages. 1978.

Vol. 651: P. W. Michor, Functors and Categories of Banach Spaces. VI, 99 pages. 1978.

Vol. 652: Differential Topology, Foliations and Gelfand-Fuks-Cohomology, Proceedings 1976. Edited by P. A. Schweitzer. XIV, 252 pages. 1978.

Vol. 653: Locally Interacting Systems and Their Application in Biology. Proceedings, 1976. Edited by R. L. Dobrushin, V. I. Kryukov and A. L. Toom. XI, 202 pages. 1978.

Vol. 654: J. P. Buhler, Icosahedral Golois Representations. III, 143 pages. 1978.

Vol. 655: R. Baeza, Quadratic Forms Over Semilocal Rings. VI, 199 pages. 1978.

Vol. 656: Probability Theory on Vector Spaces. Proceedings, 1977. Edited by A. Weron. VIII, 274 pages. 1978.

Vol. 657: Geometric Applications of Homotopy Theory I, Proceedings 1977. Edited by M. G. Barratt and M. E. Mahowald. VIII, 459 pages. 1978.

Vol. 658: Geometric Applications of Homotopy Theory II, Proceedings 1977. Edited by M. G. Barratt and M. E. Mahowald. VIII, 487 pages. 1978.

Vol. 659: Bruckner, Differentiation of Real Functions. X, 247 pages. 1978.

Vol. 660: Equations aux Dérivée Partielles. Proceedings, 1977. Edité par Pham The Lai. VI, 216 pages. 1978.

Vol. 661: P. T. Johnstone, R. Paré, R. D. Rosebrugh, D. Schumacher, R. J. Wood, and G. C. Wraith, Indexed Categories and Their Applications. VII, 260 pages. 1978.

Vol. 662: Akin, The Metric Theory of Banach Manifolds. XIX, 306 pages. 1978.

Vol. 663: J. F. Berglund, H. D. Junghenn, P. Milnes, Compact Right Topological Semigroups and Generalizations of Almost Periodicity. X, 243 pages. 1978.

Vol. 664: Algebraic and Geometric Topology, Proceedings, 1977. Edited by K. C. Millett. XI, 240 pages. 1978.

Vol. 665: Journées d'Analyse Non Linéaire. Proceedings, 1977. Edité par P. Bénilan et J. Robert. VIII, 256 pages. 1978.

Vol. 666: B. Beauzamy, Espaces d'Interpolation Réels: Topologie et Géometrie. X, 104 pages. 1978.

Vol. 667: J. Gilewicz, Approximants de Padé. XIV, 511 pages. 1978.

Vol. 668: The Structure of Attractors in Dynamical Systems. Proceedings, 1977. Edited by J. C. Martin, N. G. Markley and W. Perrizo. VI, 264 pages. 1978.

Vol. 669: Higher Set Theory. Proceedings, 1977. Edited by G. H. Müller and D. S. Scott. XII, 476 pages. 1978.

Vol. 670: Fonctions de Plusieurs Variables Complexes III, P[...]ceedings, 1977. Edité par F. Norguet. XII, 394 pages. 1978.

Vol. 671: R. T. Smythe and J. C. Wierman, First-Passage Percula[...] on the Square Lattice. VIII, 196 pages. 1978.

Vol. 672: R. L. Taylor, Stochastic Convergence of Weighted Su[...] of Random Elements in Linear Spaces. VII, 216 pages. 1978.

Vol. 673: Algebraic Topology, Proceedings 1977. Edited by P. H[...]man, R. Piccinini and D. Sjerve. VI, 278 pages. 1978.

Vol. 674: Z. Fiedorowicz and S. Priddy, Homology of Class[...] Groups Over Finite Fields and Their Associated Infinite Loop Spa[...] VI, 434 pages. 1978.

Vol. 675: J. Galambos and S. Kotz, Characterizations of Probab[...] Distributions. VIII, 169 pages. 1978.

Vol. 676: Differential Geometrical Methods in Mathematical Physic[...] Proceedings, 1977. Edited by K. Bleuler, H. R. Petry and A. Re[...] VI, 626 pages. 1978.

Vol. 677: Séminaire Bourbaki, vol. 1976/77, Exposés 489–506[...] 264 pages. 1978.

Vol. 678: D. Dacunha-Castelle, H. Heyer et B. Roynette. Ecole d'[...] de Probabilités de Saint-Flour. VII-1977. Edité par P. L. Henneq[...] IX, 379 pages. 1978.

Vol. 679: Numerical Treatment of Differential Equations in Appl[...]tions, Proceedings, 1977. Edited by R. Ansorge and W. Törnig. 163 pages. 1978.

Vol. 680: Mathematical Control Theory, Proceedings, 1977. Ed[...] by W. A. Coppel. IX, 257 pages. 1978.

Vol. 681: Séminaire de Théorie du Potentiel Paris, No. 3, Directe[...] M. Brelot, G. Choquet et J. Deny. Rédacteurs: F. Hirsch et G. Mo[...]bodzki. VII, 294 pages. 1978.

Vol. 682: G. D. James, The Representation Theory of the Symme[...] Groups. V, 156 pages. 1978.

Vol. 683: Variétés Analytiques Compactes, Proceedings, 19[...] Edité par Y. Hervier et A. Hirschowitz. V, 248 pages. 1978.

Vol. 684: E. E. Rosinger, Distributions and Nonlinear Partial [...]ferential Equations. XI, 146 pages. 1978.

Vol. 685: Knot Theory, Proceedings, 1977. Edited by J. C. Hausma[...] VII, 311 pages. 1978.

Vol. 686: Combinatorial Mathematics, Proceedings, 1977. Ed[...] by D. A. Holton and J. Seberry. IX, 353 pages. 1978.

Vol. 687: Algebraic Geometry, Proceedings, 1977. Edited by L[...] Olson. V, 244 pages. 1978.

Vol. 688: J. Dydak and J. Segal, Shape Theory. VI, 150 pages. 19[...]

Vol. 689: Cabal Seminar 76–77, Proceedings, 1976–77. Editec[...] A.S. Kechris and Y. N. Moschovakis. V, 282 pages. 1978.

Vol. 690: W. J. J. Rey, Robust Statistical Methods. VI, 128 pages. 19[...]

Vol. 691: G. Viennot, Algèbres de Lie Libres et Monoïdes Lib[...] III, 124 pages. 1978.

Vol. 692: T. Husain and S. M. Khaleelulla, Barrelledness in To[...]logical and Ordered Vector Spaces. IX, 258 pages. 1978.

Vol. 693: Hilbert Space Operators, Proceedings, 1977. Edited[...] J. M. Bachar Jr. and D. W. Hadwin. VIII, 184 pages. 1978.

Vol. 694: Séminaire Pierre Lelong – Henri Skoda (Analyse) An[...] 1976/77. VII, 334 pages. 1978.

Vol. 695: Measure Theory Applications to Stochastic Analy[...] Proceedings, 1977. Edited by G. Kallianpur and D. Kölzow. 261 pages. 1978.

Vol. 696: P. J. Feinsilver, Special Functions, Probability Semigrou[...] and Hamiltonian Flows. VI, 112 pages. 1978.

Vol. 697: Topics in Algebra, Proceedings, 1978. Edited by M. F. N[...]man. XI, 229 pages. 1978.

Vol. 698: E. Grosswald, Bessel Polynomials. XIV, 182 pages. 19[...]

Vol. 699: R. E. Greene and H.-H. Wu, Function Theory on Manifo[...] Which Possess a Pole. III, 215 pages. 1979.